EQUILIBRIUM CONSTANTS OF LIQUID–LIQUID DISTRIBUTION REACTIONS

INTRODUCTION, AND PART 1: ORGANOPHOSPHORUS EXTRACTANTS

Dedicated to the memory of Lars Gunnar Sillèn

This work was supported by the Office of Standard Reference Data (National Standard Reference Data System) of the National Bureau of Standards, Department of Commerce, United States.

INTERNATIONAL UNION OF
PURE AND APPLIED CHEMISTRY

ANALYTICAL CHEMISTRY DIVISION
COMMISSION ON EQUILIBRIUM DATA

EQUILIBRIUM CONSTANTS OF LIQUID–LIQUID DISTRIBUTION REACTIONS

INTRODUCTION, AND PART 1: ORGANOPHOSPHORUS EXTRACTANTS

Compiled by
Y. MARCUS, A. S. KERTES and E. YANIR
Department of Inorganic and Analytical Chemistry
The Hebrew University
Jerusalem

LONDON
BUTTERWORTHS

ENGLAND: BUTTERWORTH & CO. (PUBLISHERS) LTD.
LONDON: 88 Kingsway, WC2B 6AB

AUSTRALIA: BUTTERWORTHS PTY. LTD.
SYDNEY: 586 Pacific Highway, Chatswood, NSW 2067
MELBOURNE: 343 Little Collins Street, 3000
BRISBANE: 240 Queen Street, 4000

CANADA: BUTTERWORTH & CO. (CANADA) LTD.
TORONTO: 14 Curity Avenue, 374

NEW ZEALAND: BUTTERWORTHS OF NEW ZEALAND LTD.
WELLINGTON: 26–28 Waring Taylor Street, 1

SOUTH AFRICA: BUTTERWORTH & CO. (SOUTH AFRICA) (PTY) LTD.
DURBAN: 152–154 Gale Street

Published as an additional publication to

Pure and Applied Chemistry

Suggested U.D.C. number
541·123·03: 542·615

Distributed in North America by
CRANE RUSSAK & COMPANY INC.
347 Madison Avenue, New York, NY 10017

ISBN 0 408 70570 1

Printed in Great Britain by Page Bros (Norwich) Ltd, Norwich

CONTENTS

INTRODUCTION

1. General introduction

The Commission on Equilibrium Data of the Analytical Chemistry Division of the International Union of Pure and Applied Chemistry decided in 1963 to supplement the tables of *Stability Constants**, then going into a second edition, with constants for distribution equilibria of metal complexes. Extraction equilibria of both chelates and non-chelating systems were to be included. It soon transpired that the job of compiling these data was rather large, and also that the necessarily different organization of the material would make impractical the indiscriminate inclusion of these data in the Supplement to the second edition of Stability Constants, planned for publication in 1968 (later delayed to 1971). It was then decided that D. Dyrssen and H. Freiser would take care of the compilation of data for chelating systems, while Y. Marcus and A. S. Kertes would do the same for non-chelating ones. The latter authors obtained for this purpose financial assistance from the United States National Bureau of Standards, which was very helpful in advancing the work vigorously.

An examination of the literature published in 1965 revealed that about three quarters of the relevant data deal with organophosphorus extractants. This was assumed to be the case for the period between 1955 and 1967, and also earlier these extractants constituted a large portion of those reported. Thus, a compilation of these data would be a significant portion of the whole effort, and it was decided to prepare it as a separate publication, in order to avoid finished work being unduly held up.

2. Scope

Very few papers with equilibrium constants for distribution reactions involving metal complexes were published prior to 1947. The literature search for the compilation therefore generally covers the period 1947–1968 inclusive. Some more recent publications (1969) are also included. For certain extractants, however, such as ethers and esters, significant publications are known to predate 1947. For these the literature search has started at as early a date as deemed suitable in each case.

The main access to the information is through *Chemical Abstracts*, complemented by data obtained from *Nuclear Science Abstracts, Citation Index*, review articles, conference proceedings and the references of the papers scanned. Following the practice in the compilation of *Stability Constants*, it has been the policy of the compilers to attempt to read each paper in the original, after it had been located in one of the sources listed above.

---

* *Stability Constants of Metal–Ion Complexes,* Second Edition, compiled by L. G. Sillèn (inorganic ligands) and A. E. Martell (organic ligands), The Chemical Society, London, Special Publication No. 17 (1964).

It was decided that the compilation would be of greatest utility if it were limited to include only commonly used extractants, rather than try to include every system that has been reported in the literature. The aim, thus, is to obtain all the available information on a selected number of systems. The authors' judgement is hereby involved, and they will welcome any suggestions of glaring omissions. It was found that for several organophosphorus extractants, although used in practical extraction work, no equilibrium constants have been reported, nor could they readily be obtained by the compilers from the original publications. Thus, the list of extractants which are actually presented in the Tables is considerably shorter than the list which serves as the basis for the literature search. This latter list constitutes Appendix 1 of this Introduction.

Distribution equilibria involving metal ions often depend on other equilibria, involving the extractant and/or the inorganic ligand accompanying the metal ion. Thus, the partition of water, of acids and of the extractant itself are important reactions. So are also the equilibrium data on ionic dissociation of the extractant (mainly in the aqueous phase), its dimerization or further aggregation, and its association with other solvents. Therefore, in addition to the heterogeneous equilibria of the metal distribution, other heterogeneous and homogeneous equilibria which are relevant to the overall extraction reaction are also included in the compilation.

3. General arrangement

The great success of *Stability Constants* has encouraged the authors to follow its general arrangement as far as practical. The deviation from that practice is dictated by the nature of the systems dealt with in this compilation. The main difference devolves on the participation of *two* reactants associating with the metal ion: the extractant and the inorganic ligand, rather than the *one* reactant involved in the equilibria of chelate systems. In distribution equilibria, of course, it is the extractant which is of major interest, while the inorganic ligand is often of secondary importance only. Therefore, the primary classification of the present Tables is according to the extractants, and for each, after the equilibria which do not involve a ligand are dealt with, subclassification is made according to the ligands. For neutral* extractants an anionic ligand is required for charge neutralization when the metal ion is extracted. Acidic extractants usually act themselves as the anionic ligands, though under certain conditions they act as solvating extractants as well. It is then difficult to generalize about the role of the potentially involved inorganic anion. In all cases, therefore, distribution reactions are subclassified under the inorganic anion present, which belongs to the major constituent of the aqueous phase, whether or not it actually acts as a ligand.

The information given for each distribution system is the chemical reaction, the equilibrium constant, the temperature, the composition of the aqueous phase, the composition of the organic phase and the reference to the

---

* Y. Marcus and A. S. Kertes, *Ion Exchange and Solvent Extraction of Metal Complexes*, Wiley-Interscience, London (1969).

original paper. Contrary to the practice in *Stability Constants* the method of measurement is not given, since in the great majority of cases it simply is the analysis of the equilibrium phases, often radiometrically. The literature referred to follows the compilation of the data for each extractant separately.

4. Order of extractants, ligands and metal ions

Since only a limited number of extractants (about ninety) are considered in these Tables, they were ordered according to their chemical nature and the chemistry of the extraction reaction, as listed in Appendix 1, rather than according to their empirical formulae, as used in the organic ligand section of *Stability Constants*.

For each extractant, the following equilibria are listed in the order given, as far as they apply and as far as data are available.

(*a*) Distribution of the extractant.
(*b*) Dissociation of the extractant.
(*c*) Dimerization and further aggregation of the extractant.
(*d*) Reactions of the extractant with diluents and other solvents.
(*e*) Extraction of water.
(*f*) Homogeneous equilibria of the extractant with metal ions.
(*g*) Extraction of metal ions with the extractant as the sole ligand.
(*h*) Extraction of metal ions with other ligands (potential or actual).

The ligands in (*h*) are ordered according to the order established in *Stability Constants*, with organic ligands, if present (other than the extractant), following the inorganic ligands. The metal ions mentioned in (*f*), (*g*) and (*h*) follow the order established in *Stability Constants*, Inorganic Section.

5. Reactions

Distribution equilibria usually involve two reactants with the metal ion, in various proportions, and these may be in different states of protonations. Therefore, a system of subscripts and generalized symbols, such as β_2 or $^{\star}K_3$ used in *Stability Constants*, cannot cope with the complexities of these equilibria. In order to avoid uncertainties and ambiguities, the actual equilibrium reaction to which the equilibrium constant pertains has been given in each case. The following symbolism is used.

S represents the extractant if it does not dissociate ionically, otherwise it is written as HS, SA, etc. as is appropriate. The symbol used is given at the head of the table near the name of the extractant.

A represents the inorganic ligand in the solution, and is defined in each subsection of each table, near the name of the anion it represents.

M represents the metal ion appearing in the first column of the table.

Other reactants are represented by their chemical formula, except that other extractants participating in the equilibria are represented either by S′, defined in the next line below the reaction, for those extractants used only occasionally, or by their common abbreviation. These abbreviations are listed in Appendix 2. [Cf. also Y. Marcus, E. Hoffmann and A. S. Kertes, *J. Inorg. Nucl. Chem.* **33**, 853 (1971).]

A bar over chemical symbols signifies that this species is in the organic phase.

6. Equilibrium constants

Under the heading 'log *K*' are given the logarithms to base ten of the equilibrium constants for the reactions listed. The molar (mole litre^{-1}) concentration scale is understood, and exceptions are specifically noted. The following convention is used to express the limits of confidence of these entries (as estimated by the authors of the original papers). If the relative error in the constant is $\leqslant 0.07$ in log *K*, two figures are given after the decimal point; if it is >0.07 but $\leqslant 0.2$, one figure after the decimal point is given; and if it is beyond that, the two extreme limits are listed. Following *Stability Constants*, '?' denotes the authors' doubt of the values given in the source, while '(?)', used very occasionally, denotes the compilers' doubt. Its absence, naturally, does not constitute endorsement by the compilers.

In a few cases a reaction has been examined, but found to proceed only to a completely negligible extent. Then, $K \sim 0$ is used to express this finding.

In some cases the temperature variation of the equilibrium constants enabled the authors of the original papers to calculate the enthalpy and the entropy changes of the reactions quoted. These, with the Gibbs free energy changes, are given in kcal mol^{-1} for ΔH and ΔG, and cal deg^{-1} mol^{-1} for ΔS, for the reaction as written. No attempt was made to assess the precision of these data, and they are reproduced as published.

7. Temperature

Under this heading are given the temperatures in °C at which the distribution reactions were studied. 'rt' denotes room temperature, and '?' is used if the value was not found by the compilers. Unfortunately, this is the case for distribution equilibria more often than for homogeneous complex formation studies.

8. Conditions

Under the heading 'aq.' appears the composition of the aqueous phase, which sometimes is kept as a constant ionic medium, e.g. a constant molality. The salt or acid present is designated (as HA, etc., if the ligand A^- has been defined), and its molar concentration is given. If the ionic strength has been kept constant at a given value, this is given, preceded by $\mu =$. Often the composition of the aqueous phase has been varied considerably, and then the range, or the upper value of the concentration is given. In dilute aqueous solutions, the pH is given where appropriate. The figures given pertain, in every case, to the 'initial' composition of the aqueous phase, and not to its equilibrium composition. The latter is generally very difficult to derive from the published data, while the former permits, at least, reproduction of the experimental conditions to which the equilibrium data pertain.

The entries under the heading 'org.' give the relevant information for the composition of the organic phase. The diluent used is specified, and the molar concentration, or concentration range (or wt. per cent if so given by the authors), of the extractant in the diluent is given. The nature and concentration of any additives used in the organic phase are also given. The neat, undiluted extractant, when used as the organic phase, is designated by S undil. or 100% S. Again, the 'initial' rather than the equilibrium composition of the organic phase is specified.

9. References

The entries under this heading follow the usage of *Stability Constants*. Each reference is characterized by two figures denoting the year of publication, a capital letter denoting the (first) author's name, and a lower-case letter, distinguishing references with the same year and capital letter. The following conventions are used:

64D conclusions of 64D from data in 64D or other sources
64D, 67D conclusions and data of 64D and of 67D are substantially the same
64D/57D conclusions of 64D from data in 57D
/64D conclusions of compiler from data in 64D

APPENDIX 1. LIST OF EXTRACTANTS FOR WHICH THE LITERATURE WAS SEARCHED FOR DISTRIBUTION EQUILIBRIUM CONSTANTS

I Organophosphorus extractants

Phosphoric and pyrophosphoric acids

Dibutylphosphoric acid $(C_4H_9O)_2POOH$
Bis(2-ethylhexyl)phosphoric acid $(2\text{-}C_2H_5\text{–}C_6H_{12}O)_2POOH$
Diisopentylphosphoric acid $(2\text{-}CH_3\text{–}C_4H_9O)_2POOH$
Dioctylphosphoric acid $(C_8H_{17}O)_2POOH$
Butylphosphoric acid $(C_4H_9O)PO(OH)_2$
2-Ethylhexylphosphoric acid $(2\text{-}C_2H_5\text{–}C_6H_{12}O)PO(OH)_2$
Dioctylpyrophosphoric acid* $(C_8H_{17}O)P(O)(OH)OP(OH)(O)(OC_8H_{17})$
Di-*p*-chlorophenylphosphoric acid $(p\text{-}ClC_6H_4O)_2POOH$

Phosphonic and phosphinic acids

Diisopentylphosphonic acid* $(2\text{-}CH_3\text{–}C_4H_9O)(2\text{-}CH_3\text{–}C_4H_9)POOH$
Dioctyl methylenediphosphonic acid* $(C_8H_{17}O)P(O)(OH)CH_2P(OC_{18}H_{17})OOH$

Phosphates

Tri-n-butyl phosphate $(n\text{-}C_4H_9O)_3PO$
Triisobutyl phosphate $(i\text{-}C_4H_9O)_3PO$
Tris(butoxyethyl)phosphate* $(C_4H_9OC_2H_4O)_3PO$
Tri-n-octyl phosphate $(C_8H_{17}O)_3PO$
Tetrabutyl pyrophosphate* $(C_4H_9O)_2P(O)OP(O)(OC_4H_9)_2$

*Equilibrium constants searched for, but none found.

Phosphonates, phosphinates
Dibutyl butylphosphonate $(C_4H_9O)_2(C_4H_9)PO$
Dibutyl phenylphosphonate* $(C_4H_9O)_2(C_6H_5)PO$
Diisopentyl methylphosphonate $(2\text{-}CH_3\text{–}C_4H_9O)_2(CH_3)PO$
Tetrabutyl methylenediphosphonate* $(C_4H_9O)_2P(O)CH_2P(O)(OC_4H_9)_2$
Tetrahexyl methylenediphosphonate* $(C_6H_{13}O)_2P(O)CH_2P(O)(OC_6H_{13})_2$
Octyl dioctylphosphinate $(C_8H_{17}O)(C_8H_{17})_2PO$

Phosphine oxides
Tributylphosphine oxide $(C_4H_9)_3PO$
Trioctylphosphine oxide $(C_8H_{17})_3PO$
Triphenylphosphine oxide $(C_6H_5)_3PO$
Methylene bis(dihexylphosphine) oxide $CH_2[P(C_6H_{13})_2(O)]_2$
Methylene bis(di-2-ethylhexylphosphine) oxide* $CH_2[P(2\text{-}C_2H_5\text{–}C_6H_{13})_2(O)]_2$

Sections II to V to be published subsequently

II Amines, ammonium and other 'onium' compounds

Amines
Primene JMT
Amberlite LA-2
Diisononylamine $(i\text{-}C_9H_{19})_2NH$
Didecylamine $(C_{10}H_{21})_2NH$
Trioctylamine $(n\text{-}C_8H_{17})_3N$
Triisooctylamine $(i\text{-}C_8H_{17})_3N$
Trinonylamine $(C_9H_{19})_3N$
Tridodecyl (lauryl) amine $(C_{12}H_{25})_3N$
Alamine 336 $(C_{8-10}H_{17-21})_3N$

Quaternary ammonium and other 'onium' salts
Aliquat 336 $(C_{8-10}H_{17-21})_3(CH_3)N^+$
Trilaurylmethylammonium $(C_{12}H_{25})_3(CH_3)N^+$
Dimethyl octyl benzyl ammonium $(C_8H_{17})(C_6H_5CH_2)(CH_3)_2N^+$
Dimethyl hexadecyl benzyl ammonium $(C_{16}H_{33})(C_6H_5CH_2)(CH_3)_2N^+$
Tetraheptylammonium $(C_7H_{15})_4N^+$
Tetraphenylarsonium $(C_6H_5)_4As^+$
Tetraphenylphosphonium $(C_6H_5)_4P^+$

III Carboxylic and sulfonic acids, phenols

Octanoic acid (caprylic) $C_7H_{15}COOH$
Decanoic acid (capric) $C_9H_{19}COOH$

* Equilibrium constants searched for, but none found.

Dodecanoic acid (lauric) $C_{11}H_{23}COOH$
Naphthenic acid(s)
Dodecylbenzene sulfonic acid $C_{12}H_{23}C_6H_4SO_3H$
Dinonylnaphthalene sulfonic acid $(C_9H_{19})_2C_{10}H_5SO_3H$
p-Cumyl phenol *p*-(i-C_3H_7)C_6H_4–C_6H_4OH
4-sec-Butyl-2-(α-methylbenzyl)phenol
4-(sec-C_4H_9)$_2$-($C_6H_5CH(CH_3)$)C_6H_3OH

IV Inert solvents

Heptane C_7H_{16}
Octane C_8H_{18}
Decane $C_{10}H_{22}$
Kerosene $C_{12-20}H_{26-42}$
Cyclohexane cyc-C_6H_{12}
Benzene C_6H_6
Toluene $C_6H_5CH_3$
Xylene $C_6H_4(CH_3)_2$
Carbon tetrachloride CCl_4
Chloroform $CHCl_3$
1,1,2-Trichloroethane Cl_2CHCH_2Cl
Chlorobenzene C_6H_5Cl

V Monofunctional oxygen derivatives

Alcohols
n-Butanol n-C_4H_9OH
iso-Butanol i-C_4H_9OH
iso-Pentanol (isoamyl alcohol) (2-CH_3)C_4H_8OH
Benzyl alcohol $C_6H_5CH_2OH$
n-Octanol n-$C_8H_{17}OH$

Ethers
Diethyl ether $(C_2H_5)_2O$
Di-n-propyl ether (n-C_3H_7)$_2$O
Diisopropyl ether (i-C_3H_7)$_2$O
Di-n-butyl ether (n-C_4H_9)$_2$O
β,β′-Dichlorodiethyl ether $(ClC_2H_4)_2O$
Dibutylcellosolve $C_4H_9OC_2H_4OC_4H_9$
Dibutyl carbitol $(C_4H_9OC_2H_4)_2O$
Diisopropyl carbinol (i-C_3H_7)$_2$CHOH
Pentaether $C_4H_9(OC_2H_4)_4OC_4H_9$

Ketones
Diisopropyl ketone (i-C_3H_7)$_2$CO
Methyl isobutyl ketone CH_3CO–i-C_4H_9
Cyclohexanone cyc-(CH_2–)$_5$C(O)–
1-Methylcyclohexanone cyc-(CH_2–)$_4$CH(CH_3)C(O)–

Esters
Ethyl acetate $C_2H_5OCOCH_3$
n-Butyl acetate n-$C_4H_9OCOCH_3$
Isopentyl acetate i-$C_5H_{11}OCOCH_3$

Others
Triphenylarsine oxide $(C_6H_5)_3AsO$
Trioctylarsine oxide $(C_8H_{17})_3AsO$
Nitrobenzene $C_6H_5NO_2$
Nitromethane CH_3NO_2

APPENDIX 2. LIST OF ABBREVIATIONS

BDB′P	butyl dibutylphosphinate
B(DHx′PO)M′	methylene bis(dihexylphosphine oxide)
DAM′P	diisopentyl methylphosphonate
DBB′P	dibutyl butylphosphonate
DBE	dibutyl ether
D(pCl-Ph)Ph′P	di-(*p*-chlorophenyl) phenylphosphonate
DiPE	diisopropyl ethers
DPhPh′P	diphenyl phenylphosphonate
EDB′P	ethyl dibutylphosphinate
(H)iAM′P	isopentyl methylphosphonic acid
(H)DiAP	diisopentylphosphoric acid
(H)DBP	dibutylphosphoric acid
(H)DpCl-PhP	di-*p*-chlorophenylphosphoric acid
(H)DE-HP	di-(2-ethylhexyl)phosphoric acid
(H)DiPP	diisopropylphosphoric acid
(H)OPh′P	octyl phenylphosphonic acid
HxOH	hexanol
MiBC	methyl isobutyl carbitol
MiBK	methyl isobutyl ketone
OcOH	octanol
TBP	tributyl phosphate
TiBP	triisobutyl phosphate
TB′PO	tributylphosphine oxide
TH′PO	trihexylphosphine oxide
TOA	trioctylamine
TOP	trioctyl phosphate
TO′PO	trioctylphosphine oxide
TPP	tripropyl phosphate
TPhP	triphenyl phosphate

METAL ION, M	REACTION	LOG K	TEMP.	CONDITIONS, AQ.	CONDITIONS, ORG.	REF.
	Extractant Class: Phosphates					
Extractant: Tri-n-butylphosphate (TBP), S						
Dimerization						
	$2\overline{S} = \overline{S_2}$	0.39	22	—	S 0.01 in cyclohexane	68P
		0.41	22	—	S 0.01 in dodecane	68P
		0.46	22	—	S 0.01-1.0 in hexane	67P
	$2\overline{H_2O\cdot S} = \overline{(H_2O)_2S_2}$	0.95	25	H_2O	$0.1<S<0.5$ in CS_2	68R
		0.54	25	H_2O	$0.1<S<0.7$ in CCl_4	68R
	$5\overline{(H_2O)_2\cdot S_2} = \overline{(H_2O)_{10}\cdot S_{10}}$	0.53	25	H_2O	$0.5<S<1.0$ in CS_2	68R
		4.2	25	H_2O	$0.7<S<1.2$ in CCl_4	68R
Reactions with diluents and other solvents						
	$\overline{C_6H_6} + \overline{S} = \overline{S\cdot C_6H_6}$	-1.30 (mole fract.)	?	—	S 0-100% in benzene	63Rb
	$2\overline{C_6H_6} + \overline{S} = \overline{S\cdot 2C_6H_6}$	-0.7 (mole fract.)	?			63Rb
	$2\overline{CCl_4} + \overline{S} = \overline{S\cdot CCl_4}$	-1.0 (mole fract.)	?	—	S 0-100% in CCl_4	63Rb

METAL ION, M	REACTION	LOG K	TEMP.	CONDITIONS, AQ.	CONDITIONS, ORG.	REF.
Reactions with diluents and other solvents (cont.)						
	$2\overline{CCl_4} + \overline{S} = \overline{S \cdot 2CCl_4}$	-0.8 (mole fract.)	?			63Rb
	$\overline{CHCl_3} + \overline{S} = \overline{S \cdot CHCl_3}$	0.8 (mole fract.)	20	——	S<0.7 in $CHCl_3$	62Pa
	$\overline{CHCl_3} + \overline{S} = \overline{S \cdot CHCl_3}$	0.78 (mole fract.)	?	——	S 0-100% in $CHCl_3$	63Rb
	$2\overline{CHCl_3} + \overline{S} = \overline{S \cdot 2CHCl_3}$	1.20 (mole fract.)	?			63Rb
	$\overline{CDCl_3} + \overline{S} = \overline{S \cdot CDCl_3}$	0.0	?	——	S 0-1.6 $CDCl_3$	64P
	$\overline{ClCH_2CH_2Cl} + \overline{S} = \overline{S \cdot ClCH_2CH_2Cl}$	-0.22 (mole fract.)	20	——	S >.7 in $ClCH_2CH_2Cl$	62Pa
	$\overline{(H)DBP} + \overline{S} = \overline{(H)DBP \cdot S}$	1.93	25	H_2SO_4 0.1	S + (H)DBP in benzene	68La
		2.8	25	HNO_3 1.0	S<3.43 + (H)DBP 10^{-4} in CCl_4	60Da
		2.87	25	HNO_3 0.1	(H)DBP<0.5 + S in CCl_4	64D/57D, 60Da
		2.7	?	HNO_3 0.1 + $NaNO_3$ 0.9	(H)DBP<0.03 + S<0.5 in CCl_4	67K
		2.63	25	H_2SO_4 0.1	(H)DBP + S in CCl_4	66L,67L, 68La
		1.60	25	$HClO_4$ 1.0	S<0.4 in $CHCl_3$ + (H)DBP $3.3 \cdot 10^{-5}$	57D

METAL ION, M	REACTION	LOG K	TEMP.	CONDITIONS, AQ.	CONDITIONS, ORG.	REF.
Reactions with diluents and other solvents (cont.)						
		1.19	25	HNO_3 0.1	(H)DBP<0.5 + S in $CHCl_3$	64D/ 57D,60Da
		3.77	25	HNO_3 0.1	(H)DBP<0.5 + S in hexane	64D/ 57D,60Da
		2.90	25	H_2SO_4 0.1	(H)DBP + S in hexane	66L,67L, 68La
		2.83	25	HNO_3 1.0	S<0.3 + (H)DBP 10^{-5} in kerosene	59H
		4.2 to 3.1	?	HNO_3 1.0	(H)DBP 10^{-5} + TBP 0.7-0.3 in kerosene	64H/59Hc
	$\frac{1}{2}\ [(H)DBP]_2 + \overline{S} = \overline{(H)DBP\cdot S}$	-0.38	25	HCl 0.3, 0.5	(H)DBP<0.001 + S<0.3 in benzene	67M
		-0.4	25	HNO_3 0.1	(H)DBP<10^{-4} + S 0.01-3.4 in CCl_4	63B/60Da
		0.11	25	HCl 0.3, 0.5	(H)DBP<0.008 + S<0.3 in heptane	67M
		-0.05	25	HNO_3 0-15	(H)DBP 10^{-6} -0.05 + S 0.1-100% in kerosene	63B/59H
	$\overline{(H)DBP} + 2\overline{S} = \overline{(H)DBP\cdot 2S}$	3.30	25	H_2SO_4 0.1	(H)DBP<0.1 + S<1.0 in hexane	66L,67L, 68La
	$[\overline{(H)DBP}]_2 + \overline{S} = \overline{[(H)DBP]_2\cdot S}$	0.2	25	HNO_3 0.1	(H)DBP<10^{-4} + S 0.01-3.4 in CCl_4	63B/60Da

METAL ION, M	REACTION	LOG K	TEMP.	CONDITIONS, AQ.	CONDITIONS, ORG.	REF.
Reactions with diluents and other solvents (cont.)						
	$\overline{[(H)DBP]_2} + \overline{S} = \overline{[(H)DBP]_2 \cdot S}$	-0.75	25	HNO_3 0.1	(H)DBP 0.5 + S in CCl_4	64D/ 57D,60Da
		0.22	25	H_2SO_4 0.1	(H)DBP<0.1 + S 0.01-0.1 in CCl_4	67L,68La
		-0.5	25	HNO_3 0.1	(H)DBP 0.5 in $CHCl_3$	64D/ 57D,60Da
		1.08	25	HNO_3 0.1	(H)DBP<0.5 + S in hexane	64D/ 57D,60Da
		1.23	25	H_2SO_4 0.1	(H)DBP<0.1 + S 0.01-0.1 in hexane	66L,67L, 68L
		0.6	25	HNO_3 0-15	(H)DBP 10^{-6} -0.05 + S 0.1-100% in kerosene	63B/59H
	$\overline{(H)DE\text{-}HP} + \overline{S} = \overline{(H)DE\text{-}HP \cdot S}$	0.78	25	$HClO_4$ 0.2	(H)DE-HP $1.4 \cdot 10^{-3}$ + S<1.3 in benzene	66S
		1.12	25	$HClO_4$ 0.2	(H)DE-HP $1.4 \cdot 10^{-3}$ + S<1.3 in CCl_4	66S
		0.9	25	HNO_3 0.1	(H)DE-HP<$7 \cdot 10^{-4}$ + S<$2 \cdot 10$ ^{3} in toluene	68Ld
	$\frac{1}{2}\ \overline{[(H)DE\text{-}HP]_2} + \overline{S} = \overline{(H)DE\text{-}HP \cdot S}$	-0.6	25	——	S<0.5 + (H)DE-HP<0.4 in benzene	63B/59Ba
		-0.52	25	——	S<0.5 + (H)DE-HP<0.5 in CCl_4	63B/59Ba
		-0.22	25	——	S<0.5 + (H)DE-HP<0.4 in hexane	63B/59Ba

METAL ION, M	REACTION	LOG K	TEMP.	CONDITIONS, AQ.	CONDITIONS, ORG.	REF.
Reactions with diluents and other solvents (cont.)						
	$\frac{1}{2}\ \overline{[(H)DE\text{-}HP]_2} + \overline{S} = \overline{(H)DE\text{-}HP\cdot S}$	0.0	25	——	S<0.5 + (H)DE-HP<0.4 in kerosene	63B/59Ba
	$\overline{(H)DE\text{-}HP} + 2S = \overline{(H)DE\text{-}HP\cdot 2S}$	1.13	25	$HClO_4$ 0.1	S + (H)DE-HP in CCl_4	68La
		1.1	25	$HClO_4$ 0.1	(H)DE-HP<1.0 + S<0.4 in CCl_4	68Ld
		1.5	25	$HClO_4$ 0.1	(H)DE-HP<$6\cdot10^{-5}$ + S<1.3 in hexane	68Ld
		1.2	25	HNO_3 0.1	(H)DE-HP<$7\cdot10^{-4}$ + S<$2\cdot10^{-3}$ in toluene	68Ld
	$\overline{[(H)DE\text{-}HP]_2} + \overline{S} = \overline{[(H)DE\text{-}HP]_2\cdot S}$	1.17	?	HNO_3 0.05	(H)DE-HP 0.1 + S<1.0 in CCl_4	66Z
		2.92	25	$HClO_4$ 0.1	S + (H)DE-HP in CCl_4	68La,68Ld
		0.72	?	H_2O	(H)DE-HP + S undil.	69Y
	$\overline{[(H)DE\text{-}HP]_9} + \overline{S} = \overline{[(H)DE\text{-}HP]_9\cdot S}$	5.4	25	$HClO_4$ 0.1	(H)DE-HP< 1.0 + S<0.4 in CCl_4	68Ld
	$\overline{TPP} + \overline{S} = \overline{TPP\cdot S}$	0.38	25	H_2O	S0-1.5 + TPP 0-8.5 in hexane	65Da
	$\overline{[(H)OPh'P]_2} + \overline{S} = \overline{[(H)OPh']_2\cdot S}$	1.95	?	HNO_3 0.05	HS' 0.1 + S<1.0 in CCl_4	66Z
	$\overline{(H)DpCl\text{-}PhP} + \overline{S} = \overline{(H)DpCl\text{-}PhP\cdot S}$	3.28	?	$HClO_4$ 0.1	(H)DpCl-PhP<<S<0.01 in CCl_4	67Kd,68Ka

METAL ION, M	REACTION	LOG K	TEMP.	CONDITIONS, AQ.	CONDITIONS, ORG.	REF.
Reactions with diluents and other solvents (cont.)						
	$\overline{(HS')_2} + 2\overline{S} = 2\overline{HS' \cdot S}$ HS' = dinonyl naphthalene sulfonic acid	1.6	?	———	S $(1\text{-}4) \cdot 10^{-3}$ + HS' $2 \cdot 10^{-3}$ in heptane	66W
Extraction of water						
	$H_2O + \overline{S} = \overline{H_2O \cdot S}$	-0.48	25	HNO_3 0.94	S 0.55 in Amsco	67Ka/62Da
		1.15	20	H_2O	S<20% in benzene	59Va
		ΔH 3.8	6-25	H_2O	S 0-0.5 in benzene	61C
		1.22	6.0			61C
		1.15	13			61C
		1.08	20			61C
		-0.92	25	H_2O	S<0.07 in CS_2	68R
		-0.82	23	H_2O	S<0.1 in CCl_4	63W
		-0.77	19-25	H_2O or HNO_3 or H_2SO_4	S 0.1 in CCl_4	67Ka
		-0.89	20	HNO_3 1.0	S in CCl_4	67Ka/56Ab
		-0.83	25	H_2O	S<0.1 in CCl_4	68R
		-0.83	23	H_2O	S<0.1 in CCl_4	69B
		-0.36	23	H_2O	S<0.2 in dichloroethane	69B

METAL ION, M	REACTION	LOG K	TEMP.	CONDITIONS, AQ.	CONDITIONS, ORG.	REF.
Extraction of water (Cont.)						
	$H_2O + \overline{S} = \overline{H_2O \cdot S}$	1.1	25	H_2SO_4 2.5-6.0	S 7.5-10% in kerosene	61B
		ΔH-0.095	20	H_2O	S in i-octane	67Kc
		-0.92	23	H_2O	S<0.07 in i-octane	69B
		ΔH 0.14	25	H_2O	S undil.	52Ka
	$5H_2O + \overline{S} = \overline{(H_2O)_5 \cdot S}$	ΔH 0.16	20	HNO_3 dil.	S in i-octane	67Kc
	$H_2O = \overline{H_2O}$	ΔH 0.28	0-60	H_2O	H_2O 1.2 M in S undil.	64Hb
		ΔH-0.29	0-60	H_2O	H_2O 3.6 M in S undil.	64Hb
		ΔH ∿ 0.4	0-60	H_2O	S nearly dry	64Hb
	$\overline{H_2O} + \overline{S} = \overline{H_2O \cdot S}$	1.15	20	H_2O	S<1.5 in C_6H_5Br	62Pa
		0.93	20	H_2O	S<1.5 in CCl_4	62Pa
		ΔG-1.8	25		$H_2O \to 0$; $S \to 0$ in CCl_4	63K
		ΔH-2.7				
		ΔS-3				
		1.98 (mole fract.)	10	H_2O	N_{H_2O} 2.10^{-3} + N_S 0.06-1.0 in CCl_4	68N
		1.84 (mole fract.)	21			68N

METAL ION, M	REACTION	LOG K	TEMP.	CONDITIONS, AQ.	CONDITIONS, ORG.	REF.
Extraction of water (cont.)						
	$\overline{H_2O} + \overline{S} = \overline{H_2O \cdot S}$	1.70 (mole fract.)	35			68N
		1.61 (mole fract.)	45			68N
		ΔH-4.1	10-45			68N
		0.30	20	H_2O	S<1.5 in $CHCl_3$	62Pa
		0.73	20	H_2O	S<1.5 in $ClCH_2CH_2Cl$	62Pa
	$\overline{H_2O \cdot S} + \overline{S} = \overline{H_2O \cdot 2S}$	-0.95 (mole fract.)	10	H_2O	N_{H_2O} $2 \cdot 10^{-3}$ + N_S 0.06-1.0 in CCl_4	68N
		-0.89 (mole fract.)	21			68N
		-0.84 (mole fract.)	35			68N
		-0.78 (mole fract.)	45			68N
		ΔH-2.0	10-45			68N
	$\overline{S} = \overline{S \cdot H_2O}$	-0.25	20	H_2O	S 1.4 in C_6H_5Br	62Pa, 63P
		-0.50	20	H_2O	S 1.5 in CCl_4	62Pa,63P

METAL ION, M	REACTION	LOG K	TEMP.	CONDITIONS, AQ.	CONDITIONS, ORG.	REF.
Extraction of water (cont.)						
	$\overline{S} = \overline{S \cdot H_2O}$	-0.23	20	H_2O	S 1.3-1.5 in $CHCl_3$	62Pa, 63P
		-0.80	20	H_2O	S 1.5 in $ClCH_2CH_2Cl$	62Pa
		-0.2	?	H_2O	S 1.5 in $ClCH_2CH_2Cl$	63P
	$2\overline{H_2O} = \overline{(H_2O)_2}$	-1.8 to -1.5 (mole fract.)	25	H_2O	S undil.	67J
Ligand: H_2O_2						
	$H_2O_2 + \overline{S \cdot H_2O} = \overline{H_2O_2 \cdot S} + H_2O$	0.93	∿23	H_2O_2 0.6-6.41	S undil.	59T
		ΔH-2.1	∿23			59T
Ligand: $CrO_4^{=}$, $A^{=}$						
	$H_2A + 3\overline{S} = \overline{H_2A \cdot 3S}$	-0.28	?	$K_2Cr_2O_7 + H_2A$ in HNO_3 or in $HClO_4$ 1.0	S 40% in benzene	60Sb
Ligand: $Cr_2O_7^{=}$, $A^{=}$						
	$H_2A + 3\overline{S} = \overline{H_2A \cdot 3S}$	-0.32	?			60Sb

METAL ION, M	REACTION	LOG K	TEMP.	CONDITIONS, AQ.	CONDITIONS, ORG.	REF.
Ligand: TcO_4^-						
	$H^+ + A^- + \overline{S} = \overline{HA \cdot S}$	2.44	?	HCl 1.0	S in CCl_4	67Kd
Ligand: ReO_4^-						
	$H^+ + A^- + \underline{\overline{S}} = \overline{HA \cdot S}$	2.16	?	HCl 1.0	S in CCl_4	67Kd
	$H^+ + A^- + yH_2O + 2\overline{S} = \overline{HA \cdot yH_2O \cdot 2S}$	-2.8	23	HA 1.6 , 2.9	S<0.5 in CCl_4	69B
	$H^+ + A^- + 2\overline{S} = \overline{HA \cdot 2S}$	0.32 to 0.53	?	H_2SO_4 1.5-2.6	S 5-50% in kerosene	67Z
	$H^+ + A^- + 5H_2O + 3\overline{S} = \overline{HA \cdot 5H_2O \cdot 3S}$	-1.7	23	HA 1.6 , 2.9	S<0.5 in CCl_4	69B
		-0.1	23	HA 0.87	S 0.02-0.3 in i-octane	69B
	$H + A^- + yH_2O + 3\overline{S} = \overline{H^+ .3S.yH_2O} + \overline{A^-}$	-3.6	23	HA 0.01 , 0.1	S<0.3 in 1.2 dichloroethane	69B
Ligand BO_3^{-3}, A^{-3}						
	$H_3A + 2\overline{S} = \overline{H_3A \cdot 2S}$	-1.7	r.t.	H_3A $6.0 \cdot 10^{-4}$-0.15	S undil.	66Ka
		ΔH -0.5	20-70			66Ka

METAL ION, M	REACTION	LOG K	TEMP.	CONDITIONS, AQ.	CONDITIONS, ORG.	REF.
Ligand: CNS^-, A^-						
	$H^+ + A^- + \overline{S} = \overline{HA \cdot S}$	0.72 to 1.2	18	HA 1.0-0.1	S 10% in kerosene	64K
		0.90 to 1.6	18	HA 1.0-0.1	S 20%-40% in kerosene	64K
Na^+	$M^+ + A^- + xH_2O = MA \cdot xH_2O$	ΔH-8.63	20-40	MA<1.2	S undil.	59C
		ΔS-31.5				
		ΔH-9.2	20-40	MA 0.1-4.0	S undil.	62A
	$\overline{M^+} + \overline{A^-} = \overline{MA}$	4.00	20	MA 0.1-4.5	S undil.	62A,59C
		4.04	30	MA 0.1-1.2	S undil.	62A,59C
		4.48	40	MA 0.5-1.7	S undil.	62A,59C
Ca^{++}	$M^{++} + 2A^- + 4\overline{S} = \overline{MA_2 \cdot 4S}$	1.3	?	$[A^-]$0.08-2.0 + HCl 1-1.5	S 0.6-3.6 in CCl_4	68G
Sr^{++}		0.3				68G
Ba^{++}		0.0				68G

METAL ION, M	REACTION	LOG K	TEMP.	CONDITIONS, AQ.	CONDITIONS, ORG.	REF.
Ligand: CNS^-, A^- (cont.)						
Y^{+3}	$\overline{MA_3 \cdot 3S} + \overline{S} = \overline{MA_3 \cdot 4S}$	2.0	?	——	MA_3 + S 0.1-0.15 in CCl_4	69N
		2.6	?	——	MA_3 + S 0.1-0.15 in cyclohexane	69N
	$\overline{MA_3 \cdot 4S} + \overline{S} = \overline{MA_3 \cdot 5S}$	1.3	?	——	MA_3 + S 0.1-0.15 in benzene	69N
La^{+3}	$M^{+3} + 3A^- + 4\overline{S} = \overline{MA_3 \cdot 4S}$	2.4	?	NaA 0.08-0.6, pH = 3	S 0.3-3.6 in benzene	65G
	$\overline{MA_3 \cdot 4S} + \overline{S} = \overline{MA_3 \cdot 5S}$	0.6	?	——	MA_3 0.03 + S<0.1 in cyclohexane	69N
Ce^{+3}	$M^{+3} + 3A^- + 4\overline{S} = \overline{MA_3 \cdot 4S}$	2.6	?	NaA 0.08-0.6, pH = 3	S 0.3-3.6 in benzene	65G
Pr^{+3}		2.4	?			65G
Nd^{+3}		2.3	?			65G
Sm^{+3}	$M^{+3} + 3A^- + 5\overline{S} = \overline{MA_3 \cdot 5S}$	1.6	?	HCl 0.1 + $[A^-]$ 0.1-0.9	S 0.5-3.0 in benzene	67G
	$\overline{MA_3 \cdot 4S} + \overline{S} = \overline{MA_3 \cdot 5S}$	3.7	?	——	$MA_3$0.01 + S<0.15 in	69N
Eu^{+3}	$M^{+3} + 3A^- + 4\overline{S} = \overline{MA_3 \cdot 4S}$	1.44	25	NH_4A 2.0	S 0.37 in kerosene	62Y
	$M^{+3} + 3A^- + 5\overline{S} = \overline{MA_3 \cdot 5S}$	1.8	?	HCl 0.1 + $[A^-]$ 0.2-1.4	S 0.5-3.0 in benzene	67G

METAL ION, M	REACTION	LOG K	TEMP.	CONDITIONS, AQ.	CONDITIONS, ORG.	REF.
Ligand: CNS^-, A^- (cont.)						
Dy^{+3}	$\overline{MA_3\cdot 3S} + \overline{S} = \overline{MA_3\cdot 4S}$	2.1	?	——	MA_3 + S 0.1-0.15 in CCl_4	69N
		2.9	?	——	MA_3 + S 0.1-0.15 in cyclohexane	69N
Dy^{+3}	$\overline{MA_3\cdot 4S} + \overline{S} = \overline{MA_3\cdot 5S}$	1.8	?	——	MA_3 + S 0.1-0.15 in benzene	69N
Er^{+3}	$\overline{MA_3\cdot 3S} + \overline{S} = \overline{MA_3\cdot 4S}$	1.8	?	——	$MA_3<0.05$ + S 0.1-0.15 in CCl_4	69N
		2.1	?	——	$MA_3<0.05$ + S 0.1-0.15 in cyclohexane	69N
	$\overline{MA\quad 4S} + \overline{S} = \overline{MA_3\cdot 5S}$	1.0	?	——	$MA_3<0.04$ + S 0.1-0.15 in benzene	69N
Fe^{+3}	$M^{+3} + 3A^- + 3\overline{S} = \overline{MA_3\cdot 3S}$	9.0	?	$[A^-]<0.1$	S 1.0 in i-octane	64Sb
		ΔH-7.8	10-35	KA	S 1.0 in i-octane	67Kc
Ligand: NO_3^-, A^-						
	$H^+ + A^- + \overline{S} = \overline{HA\cdot S}$	-0.35	?	HA<9.0	S 0.92 in Amsco	56A
		-0.66	20	HA 0.5-3.5	S<1.0 in benzene	56F
		-1.06 to -0.44	∿20	μ 0.5-1.0 (HA)	S 5-60% in benzene	58Sa

METAL ION, M	REACTION	LOG K	TEMP.	CONDITIONS, AQ.	CONDITIONS, ORG.	REF.
Ligand: NO_3^-, A^- (cont.)						
	$H^+ + A^- + \overline{S} = \overline{HA \cdot S}$	-0.78 to -0.64	∿20	μ ∿3(HA)	S 5-60% in benzene	58Sa
		-0.66	?	HA<5.66	S<2.0 in benzene	60Fa
		-0.7	20	$(H^+, NH_4^+)A^-$ 1.2	S 1.0 in benzene	60Sc
		-0.55	25	HA<2.0	S 0.1-1.5 in benzene	65Sb
		-0.6	?	HA 0.9-2.9	S 0.1-1.0 in benzene	69S
		-0.5	∿20	μ∿0.5(HA)	S 5-60% in CCl_4	58Sa
		-1.14 to -0.6	∿20	μ∿1-3(HA)	S 5-60% in CCl_4	58Sa
		0.66	?	HA $\longrightarrow$ 0	S $\longrightarrow$ 0 in CCl_4	63N
		-1.3	?	μ=4.0 $(H^+, Na^+)A^-$	S 1.1 in CCl_4	63P
		-0.75 to -1.0	25	HA 1-4.5	S 0.5 in CCl_4	64R
		-0.8 to 0.0	20	AlA_3+ HA (μ 1-15)	S in CCl_4	64Sd
		-1.2	?	HA<4	S<0.7 in CCl_4	68A
		-0.8	?	HA 0.9-2.9	S 0.1-1.0 in CCl_4	69S
		-0.8	?	μ=4.0 $(H^+, Na^+)A^-$	S 1.1 in $CHCl_3$	63P
		-1.3	?	HA 0.9-2.9	S 0.1-1.0 in $CHCl_3$	69S

METAL ION, M	REACTION	LOG K	TEMP.	CONDITIONS, AQ.	CONDITIONS, ORG.	REF.
Ligand: NO_3^-, A^- (cont.)						
	$H^+ + A^- + \overline{S} = \overline{HA \cdot S}$	-1.0	?	μ=4.0 (H^+,Na^+)A^-	S 1.1 in dichloroethane	63P
		-1.15 to -0.6	∿20	μ∿1(HA)	S 5-60% in difluorotetra-chloroethane	58Sa
		-0.93 to -0.8	∿20	μ∿3(HA)	S 5-60% in difluorotetra-chloroethane	58Sa
		ΔG 1.40	25	HA 1	S 0.05 in n-dodecane	59S
		ΔH-0.6	25			59S
		ΔS-6.7	25			59S
		$K = 0.385 - 0.155\mu + 0.024\mu^2$	25	HA 0-0.3	S 20-30% in dodecane	59J/58Ca
		0.03+log[S] (molal scale)	25	HA 0-10 m	S 0.2 in n-hexane	59B
		-0.76	20	HA 0.5 , 1.0	S 40% in kerosene	56C
		-0.36		HA 0-13.1	S 20% in kerosene	57A/51H
		-0.84			S in kerosene	57Bc
		-0.6 to -0.89	∿20	HA 0.5-0.6	S 0.5-3.0 in kerosene	57S

METAL ION, M	REACTION	LOG K	TEMP.	CONDITIONS, AQ.	CONDITIONS, ORG.	REF.
Ligand: NO_3^-, A^- (cont.)						
	$H^+ + A^- + \overline{S} = \overline{HA \cdot S}$	-0.83			S in kerosene	58E
		-0.64	∿20	μ3.0 $H^+(A^-, ClO_4^-)$	S 2.0 in kerosene	58P
		-0.80 to -0.46	∿20	μ∿1(HA)	S 5-60% in kerosene	58Sa
		-0.82 to -0.73	∿20	μ∿3(HA)	S 5-60% in kerosene	58Sa
		-0.72	23	HA 0-11	S 0.73 in kerosene	60A
		-0.75	40			60A
		-0.76	70			60A
		-0.76	25	HA<4.0	S<0.86 in kerosene	63Sb
		-0.72	20	HA<4.0	S 20% in kerosene	64B
		ΔH-0.49	20	?	S in i-octane	67Kc
		-0.92 to -0.53	∿20	μ∿1(HA)	S 5-60% in toluene	58Sa
		-0.76 to -0.64	20	μ∿3(HA)	S 5-60% in toluene	58Sa
		-0.12 to -0.72	20	HA 0.3-7.0	S 50% in o-xylene	63R

Ligand: NO_3^-, A^- (cont.)

METAL ION, M	REACTION	LOG K	TEMP.	CONDITIONS, AQ.	CONDITIONS, ORG.	REF.
	$H^+ + A^- + \overline{S} = \overline{HA \cdot S}$	-0.2 to -0.7	20	HA 1.4-2.5 + HCl 1.0-5.0	S 50% in o-xylene	63R
		-0.2 to -0.6	20	HA 0.25-1.0 + HBr 0-2.2	S 50% in o-xylene	64A
		-0.5 to -0.7	?	HA 1-4	S 50% in o-xylene	65Ka
		-0.4 to -0.6	?	$H^+(A^-, ClO_4^-)$ 1-4	S 50% in o-xylene	65Ka
		-0.5 to 0.3	20	HA 0.25-4.0 + H_2SO_4<3.0	S 50% in o-xylene	66R
		-0.90 to -0.52	∿20	μ∿1 (HA)	S 5-60% in o,m,p-xylene	58Sa
		-0.2	∿20	μ∿0.5 (HA)	S undil.	58Sa
		-0.44	∿20	μ∿1 (HA)	S undil.	58Sa
		-0.8	∿20	μ∿3 (HA)	S undil.	58Sa
		-0.81	25	$(H^+, Li^+)A^-$ 2.0	S undil.	60Ma/58Cb
		0.30	20	HA 0.14-6.2	S undil.	64Z
	$H^+(H_2O)_4 + A^- + \overline{S} = \overline{HA(H_2O)_4 \cdot S}$	-2.3 to -1.4	25	HA 1-8	S 30% in kerosene	64E
		-1.8 to -0.8	25	HA 1.3 to 8.4	S undil.	64E

METAL ION, M	REACTION	LOG K	TEMP.	CONDITIONS, AQ.	CONDITIONS, ORG.	REF.
Ligand: NO_3^-, A^- (cont.)						
	$2H^+ + 2A^- + \overline{S} = \overline{(HA)_2 \cdot S}$	-3.6	23	HA 6-11	S 0.73 in kerosene	60A
		-3.3	40			60A
		-3.9	70			60A
		-0.1	?	HA	S in kerosene	64S
	$H^+ + A^- + \overline{HA \cdot S} = \overline{(HA)_2 \cdot S}$	-2.7	?	HA 0.4-2.5	S 1.0 in benzene	58F
	$HA + \overline{S} = \overline{HA \cdot S}$	-0.70 to 0.17 (molal scale)	25	HA 0-5	S 5%-100% in Amsco	62Da
		-0.6	25	HA<5.0	S in Amsco or kerosene	62D/55G, 56H,58Ca
		-0.75	25	HA 5.0	S 20% in CCl_4	49M
	($K_1[H^+]$ invoked)	-0.19 (+ 0.72)	20	HA<4	S 1.0 in CCl_4	59Sa
		-0.5	25	HA<5.0	S in hexane	62D/61O, 57Sb
		-0.80	20	HA 1.0-7.0	S<5% in kerosene	56Ab
		-0.25 to -0.66	?	HA 0.12-2.5	S 20% in kerosene	57R/57K

METAL ION, M	REACTION	LOG K	TEMP.	CONDITIONS, AQ.	CONDITIONS, ORG.	REF.
Ligand: NO_3^- , A^- (cont.)						
	$HA + \overline{S} = \overline{HA \cdot S}$	0.9		HA	S in kerosene	60Sd
		∿0.2	25	HA<0.5	S undil.	62D/59D
		ΔH-6.9	0-70	HA anhydrous	HA 0-4.0 in S undil.	67B
	$2HA + \overline{S} = \overline{(HA)_2 \cdot S}$	-0.1		HA	S in kerosene	60Sd
	$HA_{(gas)} + \overline{S} - \overline{HA \cdot S}$	ΔH-16.3	0-70	———	HA 0-4.0 in S undil.	67B
	$(a+b)HA(\cdot xH_2O) + y\overline{S} = \overline{aHA \cdot (yS)} + bHA(\cdot zH_2O)$	ΔH 3.80	25	HA 5.0·	S undil.	65N
	$HA + \overline{HA \cdot S} = \overline{(HA)_2 \cdot S}$	-1.4	?	HA<15.8	S 0.92 in Amsco	56A
		-1.7		HA<13.1	S 20% in kerosene	57A/51H
		ΔH-2.9	50-70	HA anhydrous	HA 3.0-4.0 in S undil.	67B
	$2HA + \overline{HA \cdot S} = \overline{(HA)_3 \cdot S}$	-1.4	20	4<HA<8.8	S 1.0 in CCl_4	59Sa
	($K_1[H^+]$ invoked)	(-1.0)				
	$\overline{H^+} + \overline{A^-} = \overline{HA}$	4.05	25	———	S 100% + HA 10^{-4}-10^{-2}	60Ha
		7.66	25	———	HA 10^{-3}-10^{-2} in S undil.	67B
		6.68	25	———	HA 10^{-3}-10^{-2}+H_2O 1.0 + S undil.	67B

METAL ION, M	REACTION	LOG K	TEMP.	CONDITIONS, AQ.	CONDITIONS, ORG.	REF.
Ligand: NO^-_3, A^- (cont.)						
	$\overline{H^+} + \overline{A^-} = \overline{HA}$	5.89	25	——	HA 10^{-3}-10^{-2}+H_2O 2.0 + S undil.	67B
		5.20	25	——	HA 10^{-3}-10^{-2}+H_2O 3.0 + S undil.	67B
		4.85	25	——	HA 10^{-3}-10^{-2}+H_2O 3.6 + S undil.	67B
	$\overline{HA} + \overline{S} = \overline{HA \cdot S}$	1.29	25	HA 4.0-7.0	S undil.	60C
		1.48	25	——	HA 3.18 + S undil.	60Ma/58Cb
Li^+	$MA + 2\overline{S} = \overline{MA \cdot 2S}$	0.2	25	MA 0.01-0.1	S 0.1 in hexane	68C
Sc^{3+}	$M^{+3} + 3A^- + 3\overline{S} = \overline{MA_3 \cdot 3S}$	0.4	?	HA 6.0	S undil.	65S
Y^{+3}		∼0.0	?	$\mu \sim 3$ +(A^-,ClO_4^-)	S undil.	61S
La^{+3}		∼0.0	?			61S
Ce^{+3}		0.0	25	(H^+,Na^+)A^- 2.0-3.61	S 0.5-3.6 in benzene	58Fa
		∼0.56	23	HA<4	S in kerosene	58H
		0.51	?	MA_3 ∼0.58	S 1.76 mole/Kg solution	67N
		ΔH-3.1	?			67N
	$MA_{3(solid)} + 3S = MA_3 \cdot 3S$	ΔH-24.8	?	——	S 1.76 mole/Kg solution	67N

METAL ION, M	REACTION	LOG K	TEMP.	CONDITIONS, AQ.	CONDITIONS, ORG.	REF.
Ligand: NO_3^-, A^- (cont.)						
Ce^{+4}	$MA_4 + 2\overline{S} = \overline{MA_4 \cdot 2S}$	ΔH-25.7	25	H_2O	S undil.	64Nb
	$M(OH)A_3 + 2\overline{S} = \overline{M(OH)A_3 \cdot 2S}$	ΔH-20.5	25			65A
	$MA_4 \cdot 5H_2O_{(solid)} + 2\overline{S} = \overline{MA_4 \cdot 2S}$	ΔH-13.4	25	——	S undil.	65A,64Nb
	$M(OH)A_3 \cdot 3H_2O_{(solid)} + 2\overline{S} = \overline{M(OH)A_3 . 2S}$	ΔH-13.4	25			65A,64Nb
Pr^{+3}	$M^{+3} + 3A^- + 3\overline{S} = \overline{MA_3 \cdot 3S}$	0.57	?	MA_3 ∿0.51	S 1.76 mole/Kg solution	67N
		ΔH-3.9	?			67N
	$MA_{3(solid)} + 3\overline{S} = \overline{MA_3 \cdot 3S}$	ΔH-26.6	?	——	S 1.76 mole/Kg solution	67N

METAL ION, M	REACTION	LOG K	TEMP.	CONDITIONS, AQ.	CONDITIONS, ORG.	REF.
Ligand: NO_3^-, A^- (cont.)						
Nd^{+3}	$M^{+3} + 3A^- + 3\overline{S} = \overline{MA_3 \cdot 3S}$	∿0.0	?	$\mu = 3$, $H^+(A^-, ClO_4^-)$	S 1.0 in CCl_4	62S
		0.62	?	MA ∿0.46	S 1.76 mole/Kg solution	67N
		ΔH-3.7	?			67N
	$MA_{3(solid)} + 3\overline{S} = \overline{MA_3 \cdot 3S}$	ΔH-24.0	?	———	S 1.76 mole/Kg solution	67N
Sm^{+3}	$M^{+3} + 3A^- + 3\overline{S} = \overline{MA_3 \cdot 3S}$	0.77	?	MA_3 ∿0.43	S 1.76 mole/Kg solution	67N
		ΔH-4.5	?			67N
	$MA_{3(solid)} + 3\overline{S} = \overline{MA_3 \cdot 3S}$	ΔH-28.2	?	———	S 1.76 mole/Kg solution	67N
Eu^{+3}	$M^{+3} + 3A + 3\overline{S} = \overline{MA_3 \cdot 3S}$	(2.9 to 2.2) -3 log $[A^-]$	18-22	HA 4.0	S $6 \cdot 10^{-3}$-0.4 in heptane	64Sc, 62Sc
Yb^{+3}		∿0.0	?	$[H^+(A^-, ClO_4^-)]$	S 1.0 in CCl_4	62S
		∿0.0	?	$[H^+(A^-, ClO_4^-)]$	S undil.	61S
Zr^{4+}	$M^{+4} + 4A^- + \overline{S} = \overline{MA_4 \cdot S}$	-0.19	∿20	$\mu = 4$ (HA)	S 0.5-1.0 in kerosene	57S
	$M^{+4} + 4A^- + 2\overline{S} = \overline{MA_4 \cdot 2S}$	-1.8	23	HA 1.0-5.2	S 6 - 90% in CCl_4	61U
		∿2.3	20-33	HA<4.0	S 19% in kerosene	58H/57Aa
		1.56	20-70	HA ⟶ 0	S 0.01-2.0 in kerosene	60Aa

METAL ION, M	REACTION	LOG K	TEMP.	CONDITIONS, AQ.	CONDITIONS, ORG.	REF.
Ligand: NO_3^-, A^- (cont.)						
Zr^{4+}	$M^{+4} + 4A^- + 2\overline{S} = \overline{MA_4 \cdot 2S}$	-5.1	25	HA 1.4	S 10-30% in kerosene	64S/57Aa, 60Aa
		-1.6 to -2.2	20	HA 5.0	MA_4 0.07-0.7; S 50% in o-xylene	62Rb,63Ra
	$M^{+4} + 4A^- + 3\overline{S} = \overline{MA_4 \cdot 3S}$	-2.3	-	HA	S in CCl_4	54M
	$MO^{++} + 2H^+ + 4A^- + 2\overline{S} = \overline{H_2MOA_4 \cdot 2S}$	0.42	25	μ 3.0 $(H^+,Na^+)A^-$	S 0.51 in CCl_4	62N
Hf^{+4}	$M^{+4} + 4A^- + 2\overline{S} = \overline{MA_4 \cdot 2S}$	-2.5 to -3.2	20	HA>5.0, MA_4 0.03-0.3	S 50% in o-xylene	62Rb,63Ra
		0.6	25	HA → 0	S 50% in o-xylene	67Kb
		-0.8 to -3.2	25	HA 0.2-2.4	S 50% in o-xylene	67Kb
		∿-3.0	10	HA 1.0	S 50% in o-xylene	67Kb
		∿-2.7	40			67Kb
		∿-2.8	55			67Kb
	$MA_4 + 2\overline{S} = \overline{MA_4 \cdot 2S}$	-0.12	20	HA 0.8-2.0	S<0.2 in kerosene	62P
Th^{+4}	$M^{+4} + 4A^- + 2\overline{S} = \overline{MA_4 \cdot 2S}$	1.56	20	HA<1.7	S<0.9 in benzene	56F,56Fc
		∿4.76	23	HA<4.0	S 19% in kerosene	58H/57Aa

METAL ION, M	REACTION	LOG K	TEMP.	CONDITIONS, AQ.	CONDITIONS, ORG.	REF.
Ligand: NO_3^-, A^- (cont.)						
Th^{+4}	$M^{+4} + 4A^- + 3\overline{S} = \overline{MA_4 \cdot 3S}$	-1.2 to -0.3	~25	HA 0-3.0	S 80% or 60% in CCl_4	51G
Pa^{+5}	$M^{+5} + 5A^- + 3\overline{S} = \overline{MA_5 \cdot 3S}$	-0.7	?	HA 1.0	S<3 in benzene	65D
		-1.2	?	HA 3.0-5.0	S<2 in benzene	65D
	$M(OH)_2^{+3} + 2H^+ + 5A^- + 2\overline{S} = \overline{MA_5 \cdot 2S} + 2H_2O$	3.7	20	HA 0.4-5.0	S 0.7-3.0 in benzene	66K
	$MA_5 + 2\overline{S} = \overline{MA_5 \cdot 2S}$	0.78	20-25	$\mu \sim 6$ (HA)	S 25% in benzene	63S
Mo^{+6}	$MO_2^{++} + 2A^- + 2\overline{S} = \overline{MA_2 \cdot 2S}$	-0.96	20	HA 0.25-1.0	S 0.4-0.5 in kerosene	67V
U^{+4}	$M^{+4} + 4A^- + 2\overline{S} = \overline{MA_4 \cdot 2S}$	1.13	25	μ = 1.75 (HA)	S 0.5-0.9 in kerosene	63Sb
		0.83	25	μ = 2.25 (HA)	S 0.5-0.9 in kerosene	63Sb
		0.65	25	μ = 2.75 (HA)	S 0.5-0.9 in kerosene	63Sb
		1.6 to 2.0	25	μ = 3.75 $(NH_4^+, H^+)A^-$	S 0.5-0.9 in kerosene	63Sb
	$\overline{MA_4} + \overline{H_2O} = \overline{M(OH)A_3} + \overline{HA}$	-1.5	20	——	MA_4 0.1 + S sat. with H_2O	65W
UO_2^{++}	$M^{++} + 2A^- + 2\overline{S} = \overline{MA_2 \cdot 2S}$	1.8-1.2	20	HA 0-0.06	S 4.5% in Amsco	56Aa
		2.23		$\mu \to 0$	S 4.5% in Amsco	56Aa
		~1.34	25	HA 0.5-6	S in Amsco	56Fa

METAL ION, M	REACTION	LOG K	TEMP.	CONDITIONS, AQ.	CONDITIONS, ORG.	REF.
Ligand: NO_3^-, A^- (cont.)						
UO_2^{++}	$M^{++} + 2A^- + 2\overline{S} = \overline{MA_2 \cdot 2S}$	1.71 (molal scale)	?	μ0.03-0.06 (MA_2)	S 1.1 in Amsco	58C
		1.44	?	HA	S in benzene	60D/ 60H,58Ha
		2.27	r.t.	MA_2 0.18	S<0.1 in benzene	63Pa
		2.10	?	MA_2 in H_2O	S 0.1 in benzene	64Na
		2.08	?	$MA_2 \to 0$	S 10% in benzene	63Rb
		2.3	?	HA 0.5	S 0.19 in benzene	64R/59Hb
		1.9 to 2.1	?	MA_2 0.2-0.5	S 20-40% in benzene	67R
		2.16 to 3.16 (mole fract.)	20	MA_2 0.1-2.4	S 1.4-1.5 in C_6H_5Br	62Pa
		2.0	?	MA_2 0.2-0.3	S 1.5 in C_6H_5Br	63P
		1.55	20	MA_2<2	S 40% in C_2H_5Br	62Pa/60P
		~0.89	25	HA 0-0.5	S 20% in CCl_4	49M
		1.08	25	μ = 0.71 (HA)	S 0.19 in CCl_4	59V
		1.29	25	$\mu \to 0$ (HA 1-13)	S 0.73 in CCl_4	61N

METAL ION, M	REACTION	LOG K	TEMP.	CONDITIONS, AQ.	CONDITIONS, ORG.	REF.
Ligand: NO_3^- , A^- (cont.)						
UO_2^{++}	$M^{++} + 2A^- + 2\overline{S} = \overline{MA_2 \cdot 2S}$	1.98	20	MA_2 0.1-2.4	S 1.5 in CCl_4	62Pa
		0.3	20	MA_2 0.2-0.5	S 1.45 in 15% CCl_4 + 45% $CHCl_3$	62Pa
		0.6	20	MA_2 0.2-0.5	S 1.45 in 30% CCl_4 + 30% $CHCl_3$	62Pa
		1.0	20	MA_2 0.2-0.5	S 1.45 in 45% CCl_4 + 15% $CHCl_3$	62Pa
		0.4	20	MA_2<2 .	S 40% in CCl_4	62Pa/60P
		1.8	?	MA_2 0.2-0.3	S 1.5 in CCl_4	63P
		1.8 to 2.2	?	HA 0.5-5.0	S 0.4-1.4 in CCl_4	63P
		1.75	r.t.	MA_2 0.18	S<0.1 in CCl_4	63Pa
		1.90	?	HA 0.5	S 0.5 in CCl_4	63Pa/60Pa
		1.77	25	MA_2 0.18	S 0.5 in CCl_4	63Pa/58B
		1.74	?	$MA_2 \rightarrow 0$	S 10% in CCl_4	63Rb
		1.68	25	MA_2	S 50-100% in CCl_4	64Aa
		1.68	?	MA_2	S 0.1 in CCl_4	64Na

METAL ION, M	REACTION	LOG K	TEMP.	CONDITIONS, AQ.	CONDITIONS, ORG.	REF.
Ligand: NO_3^-, A^- (cont.)						
UO_2^{++}	$M^{++} + 2A^- + 2\overline{S} = \overline{MA_2 \cdot 2S}$	2.06	?	MA_2<0.7	S 1.4 in CCl_4	64P
		1.7	25	HA 0.05	S 0.5 in CCl_4	64R
		1.5 to 1.7	25	HA → 0	S 0.5 in CCl_4	64R/58B
		1.70 to 1.84	20	HA 0.5-4	S 0.11-1.4 in CCl_4	64Sd
		1.6	25	MA_2 0.3-0.5 + HA	S 0.5 in CCl_4	65R
		1.6	35	MA_2 0.3-0.6 + HA	S 0.5 in CCl_4	65R
		1.4	50	MA_2 0.08-0.6 + HA	S 0.5 in CCl_4	65R
		1.3	63	MA_2 0.02-0.5 + HA	S 0.5 in CCl_4	65R
		5.28	?	HA 1.0	S 0.01 in CCl_4	67Kd
		1.60 to 1.78		MA_2 0.2-0.5	S<10-20% in CCl_4	67R
		1.23	?	HA<10	S<0.7 in CCl_4	68A
		2.37	25	MA_2	S 50-100% in CCl_4	68M,64Aa
		0.32	20	MA_2 0.5-2.2	S 1.3-1.5 in $CHCl_3$	62Pa
		0.17	20	MA_2 0.28	S 0.07-0.7 in $CHCl_3$	62Pa
		0.1	20	MA_2<2	S 40% in $CHCl_3$	62Pa/60P

METAL ION, M	REACTION	LOG K	TEMP.	CONDITIONS, AQ.	CONDITIONS, ORG.	REF.
Ligand: NO_3^-, A^- (cont.)						
UO_2^{++}	$M^{++} + 2A^- + 2\overline{S} = \overline{MA_2 \cdot 2S}$	-0.17 to 0.83	?	MA_2 0.25	S 0.08-0.25 in $CHCl_3$	63P
		0.0	?	MA_2 0.2-0.3	S 1.5 in $CHCl_3$	63P
		0.1	?	HA 0.5-4.0	S 1.1 in $CHCl_3$	63P
		0.25	?	MA_2<0.7	S 1.4 in $CHCl_3$	64P
		-0.05	?	$MA_2 \rightarrow 0$	S 10% in $CHCl_3$	63Rb
		-0.80	?	MA_2	S 0.1 in $CHCl_3$	64Na
		1.25 to 0.46		MA_2<0.7	S 1.4 in $CHCl_3$ + CCl_4	64P
		-0.15		MA_2 0.2-0.5	S 10% in $CHCl_3$	67R
		0.7 to 1.2		MA_2 0.2-0.5	S 20% in $CHCl_3$	67R
		0.08		MA_2 0.2-0.5	S 40% in $CHCl_3$	67R
		1.1		MA_2 0.2-0.5	S 80% in $CHCl_3$	67R
		1.70	?	HA 0.05	S 1.79 in dibutyl-ether	57R/57K
		1.56	20	MA_2 0.3-2.4	S 1.4-1.5 in $ClCH_2CH_2Cl$	62Pa
		1.14	20	MA_2<2	S 40% in $ClCH_2CH_2Cl$	62Pa/60P

METAL ION, M	REACTION	LOG K	TEMP.	CONDITIONS, AQ.	CONDITIONS, ORG.	REF.
Ligand: NO_3^-, A^- (cont.)						
UO_2^{++}	$M^{++} + 2A^- + 2\overline{S} = \overline{MA_2 \cdot 2S}$	1.1	?	HA 0.5-4.0 $MA_2 \rightarrow 0$	S 1.1 in $ClCH_2CCH_2Cl$	63P
		1.5	?	MA_2 0.2-0.3	S 1.5 in $ClCH_2CCH_2Cl$	63P
		ΔG-1.84	25	HA 1.0	S 0.05 in n-dodecane	59S
		ΔH-6.30	25			59S
		ΔS-15.0	25			59S
		$\log K = 0.573 - 920/T + 3.43 \cdot 10^5/T^2$	0-50			59S
		1.8	25	HA<1.0	S 18% in dodecane	61M
		2.1	?	HA 1.0	S 0.05 in dodecane	64R/59Ha
		$K = 28.392 - 14.225\mu - 2.057\mu^2 + 2.035\mu^3$	25	HA 0.3-3.0	S 20-30% in dodecane	680
		1.80	?	MA_2 0.2-3.0	S 1.5 in hexane	63P
		2.05	?	$MA_2 \rightarrow 0$	S 10% in hexane	63Rb
		2.05		MA_2 0.2-0.5	S 10% in hexane	67R

METAL ION, M	REACTION	LOG K	TEMP.	CONDITIONS, AQ.	CONDITIONS, ORG.	REF.
Ligand: NO_3^-, A^- (cont.)						
UO_2^{++}	$M^{++} + 2A^- + 2\overline{S} = \overline{MA_2 \cdot 2S}$	2.0 to 1.8		MA_2 0.2-0.5	S 20% in hexane	67R
		2.7	20-25	HA 1-6	S 3-15% in kerosene	56Fb
		1.7 to 1.9	?	HA 0.05	S 3.5-0.36 in kerosene	57R/57K
		∿2.34	20-23	HA<4.0	S 19% in kerosene	58H/57Aa
		ΔH-4.3	10-50	$(H^+,Na^+)A^-$ 0.1-1.0	S 5%-100% in kerosene or CCl_4	60N
		1.43	25	HA → 0	S 20% in kerosene or CCl_4	60N
		1.8	25	HA<1.0	S 0-1 in kerosene	61M
		0.95	?	μ 1.0 (NH_4A) + 0.05 HA	S<0.8 in kerosene	62F
		1.0 to 2.0	?	μ = 0.5-5.7 $(H^+,M^{++})A^-$	S 0.75 in kerosene	62Fa
		1.0 to 1.8	?	μ = 0.5-5.1 $(H^+,M^{++})A^-$	S 30% in kerosene	62Fa/ 58Ca
		ΔH 4.66	20-70	μ = 0.06 (HA)	S 20% in kerosene	62Ra
		ΔH 3.4	20-70	μ = 0.4 (HA)	S 20% in kerosene	62Ra
		0.9	20	HA 1.0	S 20% in kerosene	62Ra

METAL ION, M	REACTION	LOG K	TEMP.	CONDITIONS, AQ.	CONDITIONS, ORG.	REF.
Ligand: NO_3^- , A^- (cont.)						
UO_2^{++}	$M^{++} + 2A^- + 2\overline{S} = \overline{MA_2 \cdot 2S}$	0.8 to 1.6	40			62Ra
		0.6 to 1.2	70			62Ra
		1.9	20	HA 2.0	S 20% in kerosene	62Ra
		1.0 to 1.7	40			62Ra
		1.0 to 1.2	70			62Ra
		0.85	20	HA 0.5	S 20% in kerosene	62Ra
		1.4 to 1.9	40			62Ra
		0.7 to 1.2	70			62Ra
		-0.76	25	HA<4.0	S<0.86 in kerosene	63Sb
		5.14	25	HA 1-5 + MA_2<0.2	S 19% in kerosene	64S/57Aa, 57Bb
		5.3	?	HA 0.6-12	S 4.8% and 19% in kerosene	64S/58A
		5.1	25	HA 0.05-5.0	S 30% in kerosene	64S/58Ca

METAL ION, M	REACTION	LOG K	TEMP.	CONDITIONS, AQ.	CONDITIONS, ORG.	REF.
Ligand: NO_3^- , A^- (cont.)						
UO_2^{++}	$M^{++} + 2A^- + 2\overline{S} = \overline{MA_2 \cdot 2S}$	5.23	25	HA 1-5	S 30% in kerosene	64S/58Ca
		5.26	25	HA 1-6	S 5-100% in kerosene	64S/59Sb
		5.2	20	HA 0.8-2.7 + 0-6.0 $(K^+, NH_4^+, Na^+, Li^+, Al^{+3})A^-$	S 19% in kerosene	64S/60Se
		1.6 to 1.8		MA_2 0.2-0.5	S 40-80% in kerosene	67R
		ΔH-5.90	25-70	HA 0-0.3	S 30% in sat. hydrocarbon	60W,57Sa, 51C
		ΔH-6.2	20	?	S in i-octane	67Kc
		1.93	20	MA< 2	S 40% in Cl_2CCCl_2	62Pa/60P
		1.75	20	MA_2<2	S 40% in $Cl_2FCCFCl_2$	62Pa/60P
		ΔH-3.61	0-50	MA_2<3	S undil.	59Ha
		ΔS 3.24	0-50			59Ha
		0.18	20	HA 1-5 + CaA_2 0-3	S undil.	57Ba
		ΔH 6.3	?	MA_2 0.06-0.3 (mole fraction)	S undil.	64N
		1.36 to 1.59		MA_2 0.2-0.5	S undil.	67R

METAL ION, M	REACTION	LOG K	TEMP.	CONDITIONS, AQ.	CONDITIONS, ORG.	REF.
Ligand NO_3^- , A^- (cont.)						
UO_2^{++}	$M^{++} + 2A^- + 2\overline{S} = \overline{MA_2 \cdot 2S}$	1.63		MA_2	S undil.	68M
	$M^{++} + 2A^- + \overline{DAM'P} + \overline{S} = \overline{MA_2 \cdot DAM'P \cdot S}$	2.04	?	μ 1.0(NH_4A) + 0.05 HA	DAM'P<0.1 + S<0.8 in kerosene	62F
	$M^{++} + A^- + \overline{DBP^-} + \overline{S} = \overline{MA(DBP) \cdot S}$	4.80	?	HA 0.04	S<1.0 + (H)DBP 10^{-2}in Amsco	64H
	$2M^{++} + 2A^- + 2\overline{DBP^-} + \overline{2S} = \overline{[M(DBP)_2][MA_2 \cdot 2S]}$	12.52	?			64H
	$M^{++} + 3A^- + H^+ + \overline{S} = \overline{MA_2 \cdot HA \cdot S}$	∿-2.7	?	HA 1.0 , 2.0	S 20% in kerosene	57R/57K
	$MA_2 \cdot 6H_2O + 2\overline{S} = \overline{MA_2 \cdot S} + 6H_2O$	3.42	25	MA_2 0.05-2.00	S undil.	67D
	$MA_2 \cdot 6H_2O_{(solid)} + n\overline{S} = \overline{MA_2 \cdot 6H_2O \cdot nS}$	ΔH 0.80	25	——	$MA_2 \cdot 6H_2O$:S = 1:80	52Ka
	$MA_2 \cdot 2H_2O_{(solid)} + 2\overline{S} = \overline{MA_2 \cdot 2S} + 2\overline{H_2O}$	ΔH-10.4	25	——	$MA_2 \rightarrow 0$ + S undil.	64Ab
		ΔH-10.5	25	——	$[MA_2]:[S]$ = 1:80	52Ka
	$MA_{2(solid)} + 2\overline{S} = \overline{MA_2 \cdot 2S}$	ΔH-24.0	25	——	$MA_2 \rightarrow 0$ + S undil.	64Ab
	$\overline{MA_2 \cdot 2H_2O} + 4H_2O = \overline{MA_2 \cdot 6H_2O}$	ΔH 0.35	25	——	$MA_2 \cdot 2H_2O$: S = 1:80	52Ka
	$\overline{MA^+} + \overline{A^-} = \overline{MA_2}$	8.88	25	——	$MA_2<1.5 \cdot 10^{-4}$ + S undil.	61J
		10.4	25	——	MA_2 $1.5 \cdot 10^{-4}$ - 0.12 + S undil.	.61J
	$\overline{M_2A_3^+} = \overline{MA_2} + \overline{MA^+}$	-2.75	25	——	MA_2 10^{-2} + S undil.	61J

METAL ION, M	REACTION	LOG K	TEMP.	CONDITIONS, AQ.	CONDITIONS, ORG.	REF.
Ligand: NO_3^-, A^- (cont.)						
UO_2^{++}	$\overline{MA_3^-} = \overline{MA_2} + \overline{A^-}$	-3.82	25			61J
	$\overline{MA_2 \cdot 2S} + \overline{CCl_4} = \overline{MA_2 \cdot 2S \cdot CCl_4}$	K → 0	?	———	S 10% in CCl_4	63Rb
	$\overline{MA_2 \cdot 2S} + \overline{2CCl_4} = \overline{MA_2 \cdot 2S \cdot 2CCl_4}$	-0.4	?			63Rb
	$\overline{MA_2 \cdot 2S} + \overline{C_6H_6} = \overline{MA_2 \cdot 2S \cdot C_6H_6}$	-0.04	?	———	S 10% benzene	63Rb
	$\overline{MA_2 \cdot 2S} + \overline{CHCl_3} = \overline{MA_2 \cdot 2S \cdot CHCl_3}$	-0.3	?	———	S 10% in $CHCl_3$	63Rb
	$\overline{MA_2 \cdot 2S} + \overline{2CHCl_3} = \overline{MA_2 \cdot 2S \cdot 2CHCl_3}$	0.0	?	———	S 10% in $CHCl_3$	63Rb
	$\overline{M(DBP)_2} + \overline{S} = \overline{M(DBP)_2 \cdot S}$	4.95	22	HA 2.0	(H)DBP $2 \cdot 10^{-3}$ + S<0.1 in benzene	62Sa
	$\overline{M(DiAP)_2} + \overline{S} = \overline{M(DiAP)_2S}$	4.14	22	HA 2.0	(H)DiAP $3 \cdot 10^{-3}$ + S<0.1 in benzene	62Sa
		4.34	10	μ = 2 (HA)	S $6.3 \cdot 10^{-3}$ + (H)DiAP $2 \cdot 10^{-3}$ in xylene	61Sb
		3.94	60			61Sb
	$\overline{M(iAM'P)_2} + \overline{S} = \overline{M(iAM'P)_2 \cdot S}$	3.20	?	HA 2.0	S 10^{-4} - 0.1 + (H)iAM'P in benzene	62Sb
	$\overline{MA_2 \cdot 2S} + \overline{2TPP} = \overline{MA_2 \cdot 2TPP} + 2\overline{S}$	0.60	-60	———	TPP 0.2 + S 0.33 in $CDCl_3$	67S
NpO_2^{++}	$M^{++} + 2A^- + 2\overline{S} = \overline{MA_2 \cdot 2S}$	4.8	?	HA ?	S in kerosene	64S
		-0.07	20	HA 0.1-18	S undil.	64Z

METAL ION, M	REACTION	LOG K	TEMP.	CONDITIONS, AQ.	CONDITIONS, ORG.	REF.
Ligand: NO_3^- , A^- (cont.)						
Pu^{+3}	$M^{+3} + 3A^- + 2\overline{S} = \overline{MA_3 \cdot 2S}$	4.96	?	HA?	S in kerosene	64S
	$M^{+3} + 3A^- + 3\overline{S} = \overline{MA_3 \cdot 3S}$	-0.13	20	$\mu = 1.2\ (H^+, NH_4^+)A^-$	S 1.0 in benzene	60Sc
Pu^{+4}	$M^{+4} + 4A^- + 2\overline{S} = \overline{MA_4 \cdot 2S}$	0.48	25	HA>1	S in Amsco	56Fa
		3.55	?	HA 0.5	S 0.5 in CCl_4	63Pa/60Pa
		3.06	25	HA 2.0	S 0.5 in CCl_4	63Pa/60Pa
		3.10	?			64R/61Ma
		3.10	20	HA 0.5-4	S 0.37 , 1.1 in CCl_4	64Sd
		$K=-0.43+3.11 \cdot \mu^{-1}+1.79\ \mu^{-2}$	25	HA 0.6-6.0	S 20%-30% in dodecane	680
		∿5.2	23	HA<4.0	S 19% in kerosene	58H/57Aa
		1.07	25	HA<1	S 30% in kerosene	59R/58Ca
		0.6	25	HA 1.0	S 30% in kerosene	59R/58Ca
		0.15	25	HA>1	S 30% in kerosene	59R/58Ca
		2.2	25	HA<0.1	S 20% in kerosene	59R

METAL ION, M	REACTION	LOG K	TEMP.	CONDITIONS, AQ.	CONDITIONS, ORG.	REF.
Ligand: NO_3^- , A^- (cont.)						
Pu^{+4}	$M^{+4} + 4A^- + 2\overline{S} = \overline{MA_4 \cdot 2S}$	2.7 to 3.4	25	HA<5	S 20% in kerosene	59R
		0.15	25	HA>1.0 + MA_2<0.3	S 20% in kerosene	59R
		3.41	20	HA → 0	S 20% in kerosene	60M
		3.25	30			60M
		2.98	50			60M
		2.75	70			60M
		0.91 to -0.51	20	HA 0.5-4.0 + $[UO_2^{++}]$0.02-0.8	S 0.6-0.1 in kerosene	60M
		0.92 to -0.3	30			60M
		1.0 to 0.0	50			60M
		0.95 to 0.0	70			60M
		0.0 to 0.2	20-70	HA 2.0-8.0	S 0.73-0.03 in kerosene	60M
		ΔH-6.00	20-70	HA< 5	S 20% in kerosene	60M
		-0.13	20	HA 10	S 0.03 in kerosene	60M,62Ra
		-0.10	30			60M,62Ra

METAL ION, M	REACTION	LOG K	TEMP.	CONDITIONS, AQ.	CONDITIONS, ORG.	REF.
Ligand: NO_3^- , A^- (cont.)						
Pu^{+4}	$M^{+4} + 4A^- + 2\overline{S} = \overline{MA_4 \cdot 2S}$	0.0	50			60M,62Ra
		0.11	70			60M,62Ra
		3.4 to 2.7	20-70	HA → 0	S 20% in kerosene	62R
		7.8	25	HA 0.5-4.0	S 20-35% in kerosene	64S/58Ca, 57Bb,60M
		3.1 to 2.4	20-70	HA → 0	S undil.	62R
		3.0	20	HA 0.5-4.0	S undil.	64Sd
		3.30	20	HA 0.1-17	S undil.	64Z
		0.0 to 1.4	25	HA 0.5-4	S 30% in kerosene	66F/52C, 54R,58C
		-0.1 to 1.3	30	HA 0.5-4.2	S 30% in kerosene	66E/62Sd
		-0.3 to 1.0	25-30	HA 0.5-5.0	S 20% in kerosene	66E/60Sg, 60M
		-0.5 to 1.0	20-23	HA 0.4 to 4.0	S 19% in kerosene	66E/57Bb
		0.8 to 1.8	20-23	HA 0.5-4.0 + NaA 1.0-5.5	S 19% in kerosene	66E/52M
				HA 1.0-5.0 + AlA_3 0.45	S 20% in kerosene	66E/55Ga

METAL ION, M	REACTION	LOG K	TEMP.	CONDITIONS, AQ.	CONDITIONS, ORG.	REF.
Ligand: NO_3^- , A^- (cont.)						
Pu^{+4}	$MA_4 + 2\overline{S} = \overline{MA_4 \cdot 2S}$	1.48	20-25	HA 1-6	S 3-15% in kerosene	56Fb
	$\overline{M(DiAP)_4} + \overline{S} = \overline{M(DiAP)_4 \cdot S}$	2.7	10-60	μ = 2 (HA)	S 0.02 + (H)DiAP $2 \cdot 10^{-4}$ in xylene	61Sb
PuO_2^{++}	$M^{++} + 2A^- + 2\overline{S} = \overline{MA_2 \cdot 2S}$	0.97	25	HA 2.0	S 0.5 in CCl_4	63Pa/58B
		0.7 to 0.9	20	HA 0.5-4.0	S 0.11-1.1 in CCl_4	64Sd
		0.7-0.84	25	0.1<HA<5	S 20% in kerosene	59R
		0.6	25	HA<0.1	S 20% in kerosene	59R
		1.0	25	HA → 0	S 20% in kerosene	62R
		4.4	?	HA ?	S in kerosene	64S
		0.8	20	HA 0.5-4.0	S undil.	64Sd
		0.25	20	HA 0.6-14	S undil.	64Z
Am^{+3}	$M^{+3} + 3A^- + 3\overline{S} = \overline{MA_3 \cdot 3S}$	-2.18	∿25	μ = 1.0-5.0 (H^+,Na^+)A^-	S>30% in kerosene	55W
		0.4	?	HA → 0 (Na^+,H^+,A^-,ClO_4^- 2.0-5.0)	S 0.05-4 in kerosene	62Z
	$M^{+3} + 3\overline{[(H)DE\text{-}HP]_2} + \overline{S} = \overline{M[(H)(DE\text{-}HP)_2]_3 \cdot S} + 3H^+$	4.30+3log $[H^+]$	?	HA 0.05	(H)DE-HP 0.1 + S<1.0 in CCl_4	66Z

METAL ION, M	REACTION	LOG K	TEMP.	CONDITIONS, AQ.	CONDITIONS, ORG.	REF.
Ligand: NO_3^- , A^- (cont.)						
Am^{+3}	$M^{+3} + 3\overline{[(H)OPh'P]_2} + \overline{S} = \overline{M[(H)OPh'P)_2]_3 \cdot S} + 3H^+$	6.8+3log $[H^+]$	?	HA 0.05	(H)OPh'P 0.1 + S<1.0 in CCl_4	66Z
AmO_2^{++}	$M^{++} + 2\overline{[(H)DE\text{-}HP]_2} + \overline{S} = \overline{M[(H)(DE\text{-}HP)_2]_2 \cdot S} + 2H^+$	6.11+2log $[H^+]$	?	HA 0.05	(H)DE-HP 0.1 + S<0.1 in CCl_4	66Z
	$M^{++} + 2\overline{[(H)OPh'P]_2} + \overline{S} = \overline{M[(H)(OPh'P)_2]_2 \cdot S} + 2H^+$	7.3+2log $[H^+]$	?	HA 0.05	(H)OPh'P 0.1 + S<0.1 in CCl_4	66Z
Co^{++}	$M^{++} + 2A^- + 3\overline{S} = \overline{MA_2 \cdot 3S}$	-0.77-log[S] (molal scale)	25	MA_2	S undil.	59Ha
		ΔH-8.98	0-50			59Ha
		ΔS-33.7	0-50			59Ha
	$M^{++} + 2A^- + 2\overline{S} = \overline{MA \cdot 2S}$	-1.54	150	$(Li^+/Na^+/K^+),A^-$	S 0.36 in silicone	64Ha
	$MA_2 \cdot 2H_2O_{(solid)} + n\overline{S} = \overline{MA_2 \cdot 2H_2O \cdot nS}$	ΔH-1.79	25	——	S 25% in acetone	52K
		ΔH-2.00	25	——	S 50% in acetone	52K
		ΔH-3.31	25	——	S 75% in acetone	52K
		ΔH-4.26	25	——	S 95% in acetone	52K

METAL ION, M	REACTION	LOG K	TEMP.	CONDITIONS, AQ.	CONDITIONS, ORG.	REF.
Ligand: NO_3^- , A^- (cont.)						
Co^{++}	$MA_2 \cdot 2H_2O_{(solid)} + n\overline{S} = \overline{MA_2 \cdot 2H_2O \cdot nS}$	ΔH-4.74	25	———	$[MA_2]:[S]=1:80$	52K
	$MA_2 \cdot 3H_2O_{(solid)} + n\overline{S} = \overline{MA_2 \cdot 3H_2O \cdot nS}$	ΔH 0.99	25			52K
	$\overline{MA_2 \cdot 2H_2O} + 4H_2O = \overline{MA_2 \cdot 6H_2O}$	ΔH-0.26	25			52K
$Ru(NO)^{+3}$	$M^{+3} + 3A^- + 2\overline{S} = \overline{MA_3 \cdot 2S}$	5.30	3	HA 0.1-6.0	S 1.1 in kerosene	68H
		5.20	13			68H
		5.11	20			68H
		5.00	45			68H
		4.90	70			68H
		0.87	25	NaA 0.1-4.0 + HA 0.5	S 0.7 in kerosene	68H
		1.97	25	AlA_3 0.1-2.0 + 0.5 HA	S 0.7 in kerosene	68H
		ΔH-4.10	0-20	HA 0.1-6.0	S 1.1 in kerosene	68H
		ΔH-1.90	20-70			68H
	$MA_3 + 2\overline{S} = \overline{MA_3 \cdot 2S}$	2.0	20-25	HA 1-6	S 3-15% in kerosene	56Fb
	$MA_3 + 4\overline{S} = \overline{MA_3 \cdot 4S} + 4H_2O$	4.7 to 4.1	0	HA 6.0 to 3.0	S 30% in kerosene	65F
		4.2	0	HA 0.5 to 3.0	S 5-20% in kerosene	65F

METAL ION, M	REACTION	LOG K	TEMP.	CONDITIONS, AQ.	CONDITIONS, ORG.	REF.
Ligand: NO_3^- , A^- (cont.)						
$Ru(NO)^{+3}$	$MA_2OH + 2\overline{S} = \overline{MA_2OH \cdot 2S}$	0.3	25	HA 1-6	S 3-15% in kerosene	56Fb
	$MA(OH)_2 + 2\overline{S} = \overline{MA(OH)_2 \cdot 2S}$	-1.0	25	HA<3.0		56Fb
	$H^+[MA_4(H_2O)]^- + 5\overline{S} = \overline{H[MA_4(H_2O)] \cdot 5S} + 5H_2O$	5.7	0	HA 2.9-3.8	S 30% in kerosene	65F
		4.9 to 5.3	0	HA 2-3	S 10% in kerosene	65F
Cu^{++}	$M^{+2} + 2A^- + 2\overline{S} = \overline{MA_2 \cdot 2S}$	-3.7 to-4.1	?	NaA 1-5 + HA 0.1	S 20% in benzene	60S
	$M^{+2} + 2A^- + 3\overline{S} = \overline{MA_2 \cdot 3S}$	-3.1	?	NaA 5 + HA 0.1	S 20% in kerosene	60S
		3.9	?	HA 3.0	S 20% in kerosene	60S
Ligand: $SO_4^=$, $A^=$						
	$2H^+(H_2O)_4 + A^= + \overline{S} = \overline{H_2A(H_2O)_8 \cdot S}$	-1.85	19-25	H_2A 0.5	S 0.1 in CCl_4	67Ka
		-1.78	25	H_2A 3.5-7.0	S 5% in kerosene	61B
		-1.9	25	H_2A 3.5-7.0	S 2.5% in kerosene	61B
		-2.12	25	H_2A 3.5-7.0	S 1% in kerosene	61B
	$H^+ + HA^- + yH_2O + 2\overline{S} = \overline{H_2A \cdot yH_2O \cdot 2S}$ (y = 0.11-0.28)	-1.3	25	H_2A 4.5-7.7		68Na

METAL ION, M	REACTION	LOG K	TEMP.	CONDITIONS, AQ.	CONDITIONS, ORG.	REF.
Ligand: $SO_4^{=}$, $A^{=}$ (cont.)						
	$H_2A + \overline{S} = \overline{H_2A \cdot S}$	ΔH 1.1	20-50	H_2A 8-9	S 40% in cyclohexane or in S undil.	69H
	$H_2A + 2\overline{S} = \overline{H_2A \cdot 2S}$	ΔH-10.6	0-40	H_2A anhydrous	H_2A 0-2.0 + S undil.	67B
		ΔH 2.3	20-50	H_2A 6.0	S 40% in cyclohexane	69H
		ΔH 2.0	20-50	H_2A 6.0	S undil.	69H
	$H_2A + 3\overline{S} = \overline{H_2A \cdot 3S}$	ΔH 3.5	20-50	H_2A 2-4	S 40% in cyclohexane	69H
		ΔH 3.2-3.5	20-50	H_2A 2-4	S undil.	69H
	$H_2A + \overline{H_2A \cdot 2S} = \overline{(H_2A)_2 \cdot 2S}$	ΔH-3.2	40-70	H_2A anhydrous	H_2A 0-2.0 + S undil.	67B
	$\overline{H^+} + \overline{HA^-} = \overline{H_2A}$	4.05	25	———	HA 10^{-2}-10^{-4} + S undil.	60Ha
		7.85	25	———	$H_2A<10^{-2}$ + S undil.	67B
		6.89	25	———	$H_2A<10^{-2}$ + H_2O 1.0 + S undil.	67B
		5.70	25	———	$H_2A<10^{-2}$ + H_2O 2.0 + S undil.	67B
		4.89	25	———	$H_2A<10^{-2}$ + H_2O 3.0 + S undil.	67B
		4.85	25	———	$H_2A<10^{-2}$ + H_2O 3.6 + S undil.	67B
Nb^{+5}	$HMOA_2 + 2\overline{S} = \overline{HMOA_2 \cdot 2S}$	-0.2	20	H_2A 7.5-11.3	S 60% in benzene	68K

METAL ION, M	REACTION	LOG K	TEMP.	CONDITIONS, AQ.	CONDITIONS, ORG.	REF.
Ligand: $SO_4^{=}$, $A^{=}$ (cont.)						
UO_2^{++}	$M^{+2} + A^{=} + 2\overline{S} = \overline{MA_2 \cdot 2S}$	-2.0	25	H_2A 0.1	S<2.0 in hexane	68L
	$M^{+2} + 2DBP^{-} + \overline{(H)DBP} + \overline{S} = \overline{M(DBP)_2(H)DBP.S}$	12.6	25	H_2A 0.1	(H)DBP<0.06 + S<0.1 in CCl_4	60D
	$M^{+2} + 2(H)DBP + \overline{S} = \overline{M(DBP)_2 \cdot S} + 2H^{+}$	4.6	25	H_2A 0.1	(H)DBP<10^{-3} + S<2.0 in hexane	67L/66L
	$M^{+2} + 2(H)DBP + 2\overline{S} = \overline{M(DBP)_2 \cdot 2S} + 2H^{+}$	4.1 to 4.5	25	H_2A 0.1	(H)DBP<0.1 + S<2.0 in hexane	68L
	$M^{+2} + 3(H)DBP + \overline{S} = \overline{M(DBP)_2 \cdot (H)DBP \cdot S} + 2H^{+}$	8.7 to 9.1	25	H_2A 0.1	S<0.4 + (H)DBP<10^{-3} in CCl_4	68L
		7.6 to 8.1	25	H_2A 0.1	S<2.0 + (H)DBP 0.1 in hexane	67L,68L/66L
	$M^{+2} + 4(H)DBP + \overline{S} = \overline{M(DBP)_2 \cdot [(H)DBP]_2 \cdot S} + 2H^{+}$	12.3 to 13.0	25	H_2A 0.1	S<0.4 + (H)DBP<10^{-3} in CCl_4	67L,68L/66L
		11.5	25	H_2A 0.1	S<2.0 + (H)DBP 0.1 in hexane	66L,67L,68L
	$\overline{M(DBP)_2 \cdot [(H)DBP]_2} + \overline{S} = \overline{M(DBP)_2[(H)DBP]_2 \cdot S}$	∿1.9	25	H_2A 0.1	S 0-0.1 + (H)DBP 0.06 in CCl_4	63B/60D
		2.82	25	H_2A 0.1	(H)DBP<0.1 + S<0.2 in hexane	68L
	$\overline{M(DBP)_2[(H)DBP]_2} + \overline{S} = \overline{M(DBP)_2 \cdot (H)DBP \cdot S} + \overline{(H)DBP}$	-2.8 to -3.2	25	H_2A 0.1	(H)DBP<0.06 + S<0.1 in CCl_4	60D
		-2.86	25	H_2A 0.1	(H)DBP<0.1 + S<2.0 in hexane	68L

D

METAL ION, M	REACTION	LOG K	TEMP.	CONDITIONS, AQ.	CONDITIONS, ORG.	REF.
Ligand: $SO_4^{=}$, $A^{=}$ (cont.)						
UO_2^{++}	$\overline{M(DBP)_2[(H)DBP]_2} + 2\overline{S} = \overline{M(DBP)_2 \cdot 2S} + 2\overline{(H)DBP}$	-8.62	25			68L
	$M^{+2} + \overline{[(H)DE\text{-}HP]_2} + 2\overline{S} = \overline{M[DE\text{-}HP]_2 \cdot 2S} + 2H^+$	ΔH 50	10-50	H_2A 1.0	(H)DE-HP 0.1 + S 0.1 in kerosene	64Sa
		ΔH 51	10-50	H_2A 0.15	(H)DE-HP 0.1 + S 0.1 in kerosene	64Sa
	$\overline{M[(H)DE\text{-}HP]_2(DE\text{-}HP)_2} + \overline{S} = \overline{M[(H)DE\text{-}HP]_2 \cdot} \overline{\cdot [DE\text{-}HP]_2 \cdot S}$	1.20	25	H_2A 1.5	(H)DE-HP<0.5 + S<0.5 benzene	63B/59Ba
		1.60	25	H_2A 1.5	(H)DE-HP<0.5 + S<0.5 in	63B/59Ba
		1.84	25	H_2A 1.5	(H)DE-HP<0.5 + S<0.5 in CCl_4 hexane	63B/59Ba
		2.17	25	H_2A 1.5	(H)DE-HP<0.5+ S<0.5 in kerosene	63B/59Ba
Ligand: F^- , A^-						
	$HA + \overline{S} = \overline{HA \cdot S}$	-0.4 to -0.6	20	HA<1.2	S<0.4-2.4 in kerosene	64Ba
		-0.4	20	HA<3.0	S $\hat{3}$.4 in o-xylene	66Kb
	$(HA)_2 + \overline{S} = \overline{(HA)_2 \cdot S}$	-0.7	20	HA 0.5-21.0	S 3.0-3.4 in o-xylene	66Kb
	$2(HA)_2 + \overline{S} = \overline{(HA)_4 \cdot S}$	-2.4	20	HA>16.0	S 3.0-3.2 in o-xylene	66Kb
	$3(HA)_2 + 2\overline{S} = \overline{(HA)_3 \cdot 2S}$	-3.0	20	HA 16-21	S 3.2 in o-xylene	66Kb

METAL ION, M	REACTION	LOG K	TEMP.	CONDITIONS, AQ.	CONDITIONS, ORG.	REF.
Ligand: F^-, A^- (cont.)						
Ta^{+5}	$H_xMA_y + n\overline{S} = \overline{H_xMA_y \cdot nS}$	ΔH 2.41	7-60	HA 6.4	S undil.	64Nc
Ligand: Cl^-, A^-						
	$H^+ + A^- + \overline{S} = \overline{HA \cdot S}$	ΔG 6.9	25	HA → 0	S → 0 in CCl_4	63K
		ΔH 8.3	25			63K
		ΔS 5	25			63K
		-2.7	?	$(H^+,Na^+)A^-$ 0.2-2.0	S 4-100% in benzene	66La/ 65K,65L
		-1.27	?	HA 1.6-11.5	S 0.5 in benzene	68S
		-3.0	20	HA 1-10	S 0.04-1.08 in o-xylene	63A
	$H^+ + A^- + 2H_2O + \overline{S} = \overline{HA \cdot S \cdot 2H_2O}$	-2.8	20	HA 1-10	S 0.04-1.1 in o-xylene	63A
	$H^+ + A^- + 3H_2O + \overline{S} = \overline{HA \cdot S \cdot 3H_2O}$	-2.6	20			63A
		-1.0	?	1.6<HA<9	S undil.	66Kc
		-0.2 to -1.1	?	HA 3.0-8.0 + H_2SO_4 0.6-1.0	S undil.	66Kc
	$H^+ + A^- + 3.1\ H_2O + \overline{S} = \overline{HA \cdot 3.1H_2O \cdot S}$	-1.14-log[S] (molal scale)	25	HA 1-9	S undil.	59B
	$H^+ + A^- + nH_2O + \overline{S} = \overline{HA \cdot nH_2O \cdot S}$	-4.65	25	HA 7.0-9.0	S<10% in CCl_4	68Ma
	$H^+ + A^- + 2\overline{S} = \overline{HA \cdot 2S}$	-2.8	?	$(H^+,Na^+)A^-$ 0.2-2.0	S 4-100% in benzene	66La/ 65K,65L

METAL ION, M	REACTION	LOG K	TEMP.	CONDITIONS, AQ.	CONDITIONS, ORG.	REF.
Ligand: Cl^- , A^- (cont.)						
	$H^+ + A^- + 3.5H_2O + \overline{2H_2O\cdot S} = \overline{HA(H_2O)_{5.5}\cdot 2S}$	1.03	25	HA 2.0-5.2	S undil.	65Na
	$H^+ + A^- + 3\overline{S} = \overline{HA\cdot 3S}$	-3.0	?	$(H^+,Na^+)A^-$ 0.2-2.0	S 4-100% in benzene	66La/ 65K,65L
	$2H^+ + 2A^- + \overline{S} = \overline{2HA\cdot S}$	-9.5	20	HA 1-10	S 0.04-1.08 in o-xylene	63A
	$HA + n\overline{S} = \overline{HA\cdot nS}$	-2.0	21	HA 2.7-5.5	S 20% in kerosene	59I
		-1.4	21	HA 1.5-4	S 50% in kerosene	59I
		-0.3	21	HA 4.0-15.8	S undil.	59I
	$HA + \overline{S} + nH_2O = \overline{H^+\cdot nH_2O\cdot S} + \overline{A^-}$	-6.9 to -6.2	25	HA 7.0-9.0	S<0.08 in CCl_4	68Ma
	$HA_{(gas)} + \overline{S} = \overline{HA\cdot S}$	ΔH-7.5	0-30	——	HA 0-2.0 in S undil.	67B
	$\overline{H^+} + \overline{A^-} = \overline{HA}$	4.05	25	——	S 100% + HA 10^{-2}-10^{-4}	60Ha
		7.65	25	——	HA<0.01 in S undil.	67B
		6.70	25	——	HA<0.01 + H_2O 1.0 in S undil.	67B
		5.60	25	——	HA<0.01 + H_2O 2.0 in S undil.	67B
		5.00	25	——	HA<0.01 + H_2O 3.0 in S undil.	67B
		4.66	25	——	HA<0.01 + H_2O 3.6 in S undil.	67B

METAL ION, M	REACTION	LOG K	TEMP.	CONDITIONS, AQ.	CONDITIONS, ORG.	REF.
Ligand: Cl^- , A^- (cont.)						
	$\overline{HA} + \overline{S} = \overline{HA \cdot S}$	ΔG -3.5	25	——	HA → 0 + S → 0 in CCl_4	63K
		ΔH -7.4	25			63K
		ΔS -13	25			63K
Li^+	$M^+ + A^- + 1.6H_2O + \overline{S} = \overline{MA \cdot 1.6H_2O \cdot S}$	-2.95-log[S] (molal scale)	25	MA 6-13 m	S undil.	59B
Cs^+	$M^+ + A^- + \overline{S} = \overline{MA \cdot S}$	-3.64-log[S] (molal scale)	25	MA ?	S undil.	59B
Ce^{+3}	$M^{+3} + 3A^- + 3\overline{S} = \overline{MA_3 \cdot 3S}$ K_1	$\log \frac{K_1}{K_2} = 1.3$	?	HA 7-10	S or DAM'P 2.0 in xylene	68Aa
	$M^{+3} + 3A^- + \overline{3DAM'P} = \overline{MA_3 \cdot 3DAM'P}$ K_2					
Ti^{+4}	$M^{+4} + 4A^- + 4H_2O + 2\overline{S} = \overline{[H_3O \cdot S]_2{}^{++} \cdot [\cdot M(OH)_2A_4]^=}$	-1.0	?	HA 6-10	S>0.5 in benzene	68S
Zr^{+4}	$M^{+4} + 4A^- + 2\overline{S} = \overline{MA_4 \cdot 2S}$	4.3	25	$\mu = 6.54$ $H^+(A^-, ClO_4^-)$	S 0.19 in benzene	57S/56L
		-3.7 to -4.1	?	HA 5.0 + MA_4 0-0.3	S 50% in o-xylene	62Rb,63R
Hf^{+4}		-4.7 to -5.1	?	HA 5.0 + MA_4 0-0.3	S 50% in o-xylene	62Rb,63R

METAL ION, M	REACTION	LOG K	TEMP.	CONDITIONS, AQ.	CONDITIONS, ORG.	REF.
Ligand: Cl^- , A^- (cont.)						
Th^{+4}	$M^{+4} + 4A^- + 2\overline{S} = \overline{MA_4 \cdot 2S}$ K_1	$\log \frac{K_1}{K_2} = 2.8$	?	HA 6-11	S or DAM'P 0.8 in xylene	68Aa
	$M^{+4} + 4A^- + \overline{2DAM'P} = \overline{MA_4 \cdot 2DAM'P}$ K_2					
Nb^{+5}	$M(V) + 4A^- + 3\overline{S} = \overline{(H_3O\ S)_2{}^{++} \cdot}$ $\overline{\cdot [M(OH)_3A_4 \cdot S]^=}$	-1.8	?	HA 8-10	S 0.1 in benzene	68Sa
U^{+4}	$M^{+4} + 4A^- + 2\overline{S} = \overline{MA_4 \cdot 2S}$ K_1	$\log \frac{K_2}{K_1} \sim 2.5$	?	HA 5.5-6.5	S or DAM'P 20% in CCl_4	61Sa
	$M^{+4} + 4A^- + \overline{2DAM'P} = \overline{MA_4 \cdot 2DAM'P}$ K_2					
	$\overline{MA_4 \cdot 2S} + \overline{S} = \overline{MA_4 \cdot 3S}$	1.65	?	——	S<1.0 in CCl_4	64L
		1.74	?	——	S 20% in CCl_4	67Va
		1.74	?	——	S 0-0.3 in kerosene	65La
$UO_2{}^{++}$	$M^{++} + 2A^- + 2\overline{S} = \overline{MA_2 \cdot 2S}$ K_1	$\log \frac{K_2}{K_1} = 2.07$	?	HA 3-5.5	S or DAM'P 20% in CCl_4	61Sa
	$M^{++} + 2A^- + \overline{2DAM'P} = \overline{MA_2 \cdot 2DAM'P}$ K_2					
	$\overline{MA_2 \cdot 2S} + \overline{S} = \overline{MA_2 \cdot 3S}$	0.5	?	——	MA_2<0.03 + S 0-3.67 in CCl_4	63V,67Va

METAL ION, M	REACTION	LOG K	TEMP.	CONDITIONS, AQ.	CONDITIONS, ORG.	REF.
Ligand: Cl^- , A^- (cont.)						
Pu^{+4}	$M^{+4} + 4A^- + 2\overline{S} = \overline{MA_4 \cdot 2S}$ K_1	$\log \frac{K_2}{K_1} = 0.34$	?	HA 2.5-4.5	TOA HA 0.047 + S 0.76 in 0-xylene	60Sf
	$M^{+4} + 4A^- + \overline{2TOA \cdot HA} = \overline{MA_4 \cdot TOA \cdot HA}$ K_2					
Os^{+4}	$H_2MA_6 + 4\overline{S} = \overline{H_2MA_6 \cdot 4S}$	ΔH-18.45	0-20	HA 3.0	S 20% in dibutylether	69M
		ΔH-6.45	0-50	HA 6.0	S 20% in dibutylether	69M
		ΔH-5.54	0-50	HA 9.0	S 20% in dibutylether	69M
Ag^+	$M^+ + H^+ + 2A^- + 3\overline{S} = \overline{HMA_2 \cdot 3S}$	4.6 to 5.7		$(H^+,Na^+)A^-$ 0-5.0	S 4-100% in benzene	65K,65L, 66L
Au^{+3}	$H^+ + M^{+3} + 4A^- + 3\overline{S} = \overline{HMA_4 \cdot 3S}$	-0.23	?	?	S<0.1 in i-octane	64Sb
Zn^{++}	$M^{++} + 2A^- + 2\overline{S} = \overline{MA_2 \cdot 2S}$	-1.9	?	?	S in i-octane	64Sb
	$MA_3^- + H^+ + 3\overline{S} = \overline{HMA_3 \cdot 3S}$	0.30	20	HA 0.83-2.7	S undil.	62M
	$MA_4^{-2} + 2H^+ + 2\overline{S} = \overline{H_2MA_4 \cdot 2S}$	-0.4	20	HA>0.83	S undil.	62M
	$MA_2 + 2\overline{S} = \overline{MA_2 \cdot 2S}$	-0.8	20	HA<0.83	S undil.	62M
Hg^{++}	$M^{++} + 2A^- + 2\overline{S} = \overline{MA_2 \cdot S}$	5.7	?	?	S in i-octane	64Sb
Tl^{+3}	$MA_3 + 2\overline{S} = \overline{MA_3 \cdot 2S}$	ΔH 8.5	13-80	H_2SO_4 0.5 + NaA 0.007	S 3% in decane:octane 1:1	65Kb
Te^{+4}	$MA_4(H_2O)_2 + 2\overline{S} = \overline{MA_4 \cdot 2S} + 2H_2O$	0.0	18	HA 4.5-10.7	S 0.12-0.4 in i-octane	68Sb
	$MO_2H^+ + 3H^+ + 4A^- + 3\overline{S} = \overline{MA_4 \cdot 3S} + 2H_2O$	-2.80	22	HA 1.5-3.0	S 1.82 in n-hexane	65B

METAL ION, M	REACTION	LOG K	TEMP.	CONDITIONS, AQ.	CONDITIONS, ORG.	REF.
Ligand: Cl^- , A^- (cont.)						
Po^{+4}	$MA_5^- + 2\overline{S} = \overline{MA_4 \cdot 2S} + A^-$	0.8	?	HA 0.1-0.5	S<0.3 in benzene	65Sa
	$MA_5^- + H^+ + 2\overline{S} = \overline{HMA_5 \cdot 2S}$	1.8	?	HA 4-6	S<0.3 in benzene	65Sa
	$MA_4 + 2\overline{S} = \overline{MA_4 \cdot 2S}$	-1.4	22	HA 6-8	S 0.04-0.07 in decaline	57B
Ligand: ClO_4^- , A^-						
	$H^+ + A^- + \overline{S} = \overline{HA \cdot S}$	-0.75 to -1.35	∿20	HA 0.9-9.0	S 1.0 in benzene	58P
		-0.7 to 0.0	∿20	μ ∿ 0.5 (HA)	S 1.5-3.6 in benzene	58S
		-0.5 to 1.0	∿20	μ ∿ 1 (HA)	S 1.5-3.6 in benzene	58S
		-1.04	∿20	μ ∿ 2 (HA)	S 1.5-3.7 in benzene	58S
		-1.2	∿20	μ ∿ 3 (HA)	S 0.8-3.7 in benzene	58S
		-1.18	∿20	HA 3-4	S 0.8-3.6 in benzene	58S
		-1.48 to 0.09	∿20	HA 0.5-8.9	S 2-3.6 in kerosene	58P
		-1.8 to -1.5	?	$7 > H^+(NO_3^-, A^-) > 4.5$	S 50% in o-xylene	65Ka
	$H^+ + A^- + 5H_2O + 3\overline{S} = \overline{HA \cdot 5H_2O \cdot 3S}$	-0.1	23	HA<10	S 0.07, 0.37 in i-octane	69B
	$H^+ + A^- + yH_2O + 2\overline{S} = \overline{HA \cdot yH_2O \cdot 2S}$	-2.3	23			69B

METAL ION, M	REACTION	LOG K	TEMP.	CONDITIONS, AQ.	CONDITIONS, ORG.	REF.
Ligand: ClO_4^- , A^-						
	$H^+ + A^- + 4\overline{S} = \overline{HA\cdot 4S}$	-0.8	∿20	HA 0.01-3.17	S 0.5-2.5 in benzene	60F
		-0.4	∿20	HA 10	S 0.5-1.5 in benzene	60F
		-0.6	?	HA 0.6-10	S in benzene	64Se
	$2H^+ + 2A^- + 4\overline{S} = \overline{2HA\cdot 4S}$	-2.67	?			64Se
	$3H^+ + 3A^- + 4\overline{S} = \overline{3HA\cdot 4S}$	-4.85	?			64Se
	$4H^+ + 4A^- + 4\overline{S} = \overline{4HA\cdot 4S}$	-7.0	?			64Se
Ca^{++}	$M^{++} + 2A^- + 4\overline{S} = \overline{MA_2\cdot 4S}$	2.3-4 log [S]	?	HA 0.1-3.0	S 0.2-3.6 in benzene	68Lb
Sr^{++}		-1.85-4 log[S]	?			68Lb
Ba^{++}		-1.62-4 log[S]	?			68Lb
La^{+3}	$M^{+3} + 3A^- + 4\overline{S} = \overline{MA_3\cdot 4S}$	-0.4-4 log[S]	?	$[H^+,Na^+]A^-$ 1.6-2.0	S 0.3-3.6 in benzene	64G
Ce^{+3}		-0.1-4 log[S]	?			64G
Pr^{+3}		0.0-4 log [S]	?			64G
Nd^{+3}		-0.4-4 log[S]	?			64G

METAL ION, M	REACTION	LOG K	TEMP.	CONDITIONS, AQ.	CONDITIONS, ORG.	REF.
Ligand: Br^- , A^-						
	$H^+ + A^- + \overline{S} = \overline{HA\cdot S}$	1.16	?	$(H^+,Na^+)A^-$ 0.5-4.0	S 2.0-3.4 in benzene	66La
	$H^+ + A^- + 3.6H_2O + \overline{S} = \overline{HA\cdot 3.6\ H_2O\cdot S}$	-0.50-log[S] (molal scale)	25	HA 1-10 m	S undil.	59B
Li^+	$M^+ + A^- + 2.5H_2O + \overline{S} = \overline{MA\cdot 2.5H_2O\cdot S}$	-1.98-log[S] (molal scale)	25	MA 3-12 m	S undil.	59B
Ag^+	$M^+ + H^+ + 2A^- + 3\overline{S} = \overline{HMA_2\cdot 3S}$	9.8	?	$(H^+,Na^+)A^-$ 0.2-4.0	S 1.0-3.4 in benzene	66La
Au^{+3}	$M^+ + H^+ + 4A^- + 3\overline{S} = \overline{HMA_4\cdot 3S}$	4.28	?	?	S<0.1 in i octane	64Sb
Zn^{++}	$M^{++} + 2A^- + 2\overline{S} = \overline{MA_2\cdot 2S}$	-0.9	?	?	S in i-octane	64Sb
	$MA_3^- + H^+ + 3\overline{S} = \overline{HMA_3\cdot 3S}$	1.08	20	HA>0.8	S undil.	62M
	$MA_4^= + 2H^+ + 2\overline{S} = \overline{H_2MA_4\cdot 2S}$	1.0	20	HA<2	S undil.	62M
	$MA_2 + 2\overline{S} = \overline{MA_2\cdot 2S}$	0.23	20	HA<0.8	S undil.	62M
Hg^{++}	$M^{++} + 2A^- + 2\overline{S} = \overline{MA_2\cdot 2S}$	9.5	?		S in i-octane	64Sb
	$Li^+ + M^{++} + 3A^- + \overline{S} = \overline{Li\ MA_3\cdot S}$	7.9	?		S in i-octane	64Sb
Ligand: I^- , A^-						
	$H^+ + A^- + 3.9H_2O + \overline{S} = \overline{HA\cdot 3.9H_2O\cdot S}$	0.85-log[S] (molal scale)	25	HA 0-8 m	S undil.	59B

METAL ION, M	REACTION	LOG K	TEMP.	CONDITIONS, AQ.	CONDITIONS, ORG.	REF.
Ligand: I^- , A^- (cont.)						
Li^+	$M^+ + A^- + H_2O + 3\overline{S} = \overline{MA \cdot 3.6H_2O \cdot 3S}$	-3.53	?	MA 3-6	S 0.5 in benzene	67Ra
		-0.54-log[S] (molal scale)	25	MA 0.5-2.0 m	S undil.	59B
Na^+	$M^+ + A^- + 4.9H_2O + \overline{S} = \overline{MA \cdot 4.9H_2O \cdot S}$	-1.32-log[S] (molal scale)	25	MA 2-3.5 m	S undil.	59B
K^+	$M^+ + A^- + 3.7H_2O + \overline{S} = \overline{MA \cdot 3.7H_2O \cdot S}$	-1.64-log[S] (molal scale)	25	MA 2-5 m	S undil.	59B
Ca^{++}	$M^{++} + 2A^- + 4\overline{S} = \overline{MA_2 \cdot 4S}$	0.39-4log[S]	?	NaA 0.4-3.2 + 0.5 HCl	S undil.	68Lc
Sr^{++}		0.12-4log[S]	?			68Lc
Ba^{++}		-0.08-4log[S]	?			68Lc
Zn^{++}	$M^{++} + 2A^- + 2\overline{S} = \overline{MA_2 \cdot 2S}$	0.2	?	?	S in i-octane	64Sb
		ΔH-4.7	20	LiA	S in i-octane	67Kc
Cd^{++}		2.8	?	$[A^-]<0.3$ $[M^{++}]$ 0.1	S undil.	64Sb
		ΔH-5.5	20	LiA	S in i-octane	67Kc
Hg^{++}		11.8	?	?	S in i-octane	64Sb
	$Li^+ + M^{++} + 3A^- + \overline{S} = \overline{LiMA_3 \cdot S}$	9.4	?		S in i-octane	64Sb

METAL ION, M	REACTION	LOG K	TEMP.	CONDITIONS, AQ.	CONDITIONS, ORG.	REF.
Ligand: I^- , A^- (cont.)						
Hg^{++}	$M^{++} + 2A^- + 2\overline{S} = \overline{MA_2 \cdot 2S}$	ΔH-8.0	10-35	LiA	S in i-octane	67Kc
	by calorimetric method	ΔH-10.1	20			67Kc
Bi^{+3}	$M^{+3} + 3A^- + 3\overline{S} = \overline{MA_3 \cdot 3S}$	ΔH-4.8	10-35	LiA	S 1.0 in i-octane	67Kc
		ΔS 13.2	10-35			67Kc
	$\overline{A^+} + \overline{A_3^-} = \overline{2A_2}$	8.2	25	——	A_2 0.01 + S 1.0 in n-octane	64V
		3.5 to 3.2	25	——	$A_2 10^{-3}$-0.02 in S undil.	64V
	$A_2 + \overline{S} = \overline{A_2 \cdot S}$	1.40	10	——	$S<0.12 + A_2\ 6.8 \cdot 10^{-4}$ in heptane	60T
		1.34	17			60T
		1.29	25			60T
		1.22	35			60T
		ΔH-2.94	10-35	——	$S\ 0.12 + A_2\ 6.8\ 10^{-4}$ in heptane	60T
		ΔS-3.99	10-35			60T

METAL ION, M	REACTION	LOG K	TEMP.	CONDITIONS, AQ.	CONDITIONS, ORG.	REF.
Ligand: Organic acids						
	$C_4H_6O_5 + 1.5H_2O + 2\overline{S} = \overline{C_4H_6O_5 \cdot 1.5H_2O \cdot 2S}$	-0.96	?	$C_4H_6O_5$ 0.56-2.63	S 0.55-3.67 in n-dodecane	63Sa
	(malic acid)	-0.96	?	$\mu \to 0$ $C_4H_6O_5$ 0.5-3.2	S 60% in n-dodecane	63Sa
		-1.01	?	μ 0.5 ($LiNO_3$) $C_4H_6O_5$ 0.2-2.5	S 60% in n-dodecane	63Sa
		-0.85	?	μ 2.0 ($LiNO_3$) $C_4H_6O_5$ 0.2-2.5	S 60% in n-dodecane	63Sa
	$C_4H_6O_6 + H_2O + 2\overline{S} = \overline{C_4H_6O_6 \cdot H_2O \cdot 2S}$	-1.4	?	$C_4H_6O_6$ 0.06-1.0	S 0.7-3.7 in n-dodecane	63Sa
	(tartaric acid)	-1.45	?	$\mu \to 0$, ($C_4H_6O_6$ 0.06-1.3)	S 60% in n-dodecane	63Sa
		-1.39	?	μ 0.5 ($LiNO_3$), $C_4H_6O_6$ 0.2-2.6	S 60% in n-dodecane	63Sa
		-1.32	?	μ 2.0 ($LiNO_3$), $C_4H_6O_6$ 0.2-2.6	S 60% in n-dodecane	63Sa
	$C_5H_8O_7 + 2\overline{S} + \overline{C_5H_8O_7 \cdot 2S}$	-1.80	?	$C_5H_8O_7$ 0.06-1.93	S 0.73-3.67 in kerosene	60Sa
	(trihydroxyglutaric acid)	-1.84	?	$\mu \to 0$ $C_5H_8O_7$ 0.1-2	S 60% in dodecane	63Sa
		-1.72	?	μ 0.5 ($LiNO_3$) $C_5H_8O_7$ 0.2-2.2	S 60% in n-dodecane	63Sa
		-1.57	?	μ 2.0 ($LiNO_3$) $C_5H_8O_7$ 0.2-2.3	S 60% in n-dodecane	63Sa
	$C_3H_6O_3 + xH_2O + \overline{S} = \overline{C_3H_6O_3 \cdot xH_2O \cdot S}$	-0.59	?	$C_3H_6O_3$ 0.8-3.2	S 0.73-3.62 in n-dodecane	63Sa
	(lactic acid)	-0.54	?	$\mu \to 0$ $C_3H_6O_3$ 0.5-2.5	S 60% in n-dodecane	63Sa
		-0.66	?	μ 0.5 ($LiNO_3$) $C_3H_6O_3$ 0.2-2.1	S 60% in n-dodecane	63Sa
		-0.57	?	μ 2.0 ($LiNO_3$) $C_3H_6O_3$ 0.2-2.0	S 60% in n-dodecane	63Sa

REFERENCES

49M Moore, R. L., USAEC Report AECD 3196 (1949)

51G Gercke, R.H., USAEC Report UCRL 1493 (1951)

51H Hardwick, W. H., Mills, A. L. and Stubbin, P., British Report AERE-CM-105 (1951) (quoted in 57A)

51C Codding, J. W. et al, USAEC Report KAPL - 602 (1951)

52C Chemical Technology Division, Oak Ridge National Laboratory, Annual Report, ORNL - 1141 (1952)

52K Katzin, L. I. and Ferraro, J. R., J. Amer. Chem. Soc., 74, 6040 (1952)

52Ka Katzin, L. I., Simon, D. M. and Ferraro, J. R., J. Amer. Chem. Soc., 74, 1191 (1952)

52M McKay, H. A. C. and Woodgate, P. R., British Report AERE - C/R - 994 (1952)

54M Murbach, E. W. and McVey, W. H., USAEC Report IRL - 115 (1954) (quoted in 61U)

54R Annual Report, USAEC Report KAPL - 1002 (1954)

55G Geary, T. V., British Report AEAR 8142 (1955)

55Ga Goulden, P. D., Durham, R. W. and Corriveau, V., USAEC Report CRDC - 640 (1955)

55W Walsh, K. A., USAEC Report LA 1861 (1955)

56A Allen, R. J. and Desesa, M. A., USAEC Report WIN - 52 (1956); Nucleonics, 15, 88 (1957)

56Aa Andelin, R. L., Anderson, E. L., McVey, H. W., USAEC Report IDO 14381 (1956)

56Ab Alcock, K., Grimley, S. S., Healy, T. V., Kennedy, J. and McKay, H. A. C., Trans. Faraday Soc., 52, 39 (1956)

56C Cacciari, A., De Leone, R., Fizzotti, C. and Cabaglio, M., Energia Nucleare, 3, 368 (1956)

56F Fomin, V. V. and Maiorova, E. P., Zh. Neorg. Khim., 1, 1703 (1956)

56Fa Flanary, J. R., Proc. Int. Conf. Peaceful Uses At. Energ. Geneva, 1955, Vol. 9 p. 528, United Nations, Geneva, 1956; "Progress in Nucl. Energy" Ser. III, Proc. Chem., Pergamon, London, 1956, Vol. 1 p. 195

56Fb Fletcher, J. M., in "Progress in Nucl. Energy", Ser. III, Proc. Chem., Vol. 1 p. 105 (1956)

56Fc Fomin, V. V. and Maiorova, E. P., Zh. Neorg. Khim., 1, 2749 (1956)

56H Healy, T. V. and Brown, P. E., British Report AERE C/R 1970 (1956)

56L Levitt, A. E. and Freund, H., J. Amer. Chem. Soc., 78, 1545 (1956)

57A Allen, R. J. and Desesa, M. A., Nucleonics, 15, 88 (1957)

57Aa Alcock, K., Bedford, F. C., Hardwick, W. H. and McKay, H. A. C., J. Inorg. Nucl. Chem., 4, 100 (1957)

57B Bagnall, K. W. and Robertson, D. S., J. Chem., Soc., 1957, 509 (1957)

57Ba Bernstrom, B. and Rydberg, J., Acta Chem. Scand., 11, 1173 (1957)

57Bc Benedict, M. and Pifgord, T. H., Quoted in 64B

57Bb Best, G., McKay, H. A. C. and Woodgate, P., J. Inorg. Nucl. Chem., 4, 315 (1957)

57D Dyrssen, D., Acta Chem. Scand., 11, 1771 (1957)

57K Karpacheva, S. M., Khorkhorina, L. P. and Rozen, A. M., Zh. Neorg. Khim., 2, 1441 (1957)

57R Rozen, A. M., Zh. Neorg. Khim., 2, 1956 (1957)

57S Solovkin, A. S., Zh. Neorg. Khim., 2, 611 (1957)

57Sa Siddall, T. H., et al, USAEC Report DP - 53 (1957)

57Sb Savolainen, J. R., in USAEC Report ORNL - 1367 (1957)

58A Alcock, K., Best, G., Hesford, E. and McKay, H. A. C., J. Inorg. Nucl. Chem., 6, 328 (1958)

58B Burger, L. L., J. Phys. Chem., 62, 590 (1958)

58C Codding, J. W. Jr., USAEC Report IDO - 14454 (1958)

58Ca Codding, J. W., Haas, W. O. and Heumann, F. K., Ind. Eng. Chem., 50, 145 (1958)

58Cb Collopy, T. J., USAEC Report NCLO - 749 (1958)

58E Eiichi, I., et al., Rikagaku Kenkyssho Hokoku, 34, 276 (1958) (quoted in 64B)

58F Fomin, V. V. and Maiorova, E. P., Zh. Neorg. Khim., 3, 540 (1958)

58Fa Fomin, V. V. Kartushova, R. E. and Rudenko, T. I., Zh. Neorg. Khim., 3, 2117 (1958)

58H Hesford, E. and McKay, H. A. C., Trans. Faraday Soc., 54, 573 (1958)

58Ha Healy, T. V. and Kennedy, J. British Report AERE C/R 2686 (1958) (quoted in 60D)

58P Povitskii, N. S., Solovkin, A. S. and Shilin, I. V., Zh. Neorg. Khim., 3, 222 (1958)

58S Shevchenko, V. B., Shilin, I. V. and Solovkin, A. S., Zh. Neorg. Khim., 3, 225 (1958)

58Sa Shevchenko, V. B., Povitskii, N. S., Solovkin, A. S., Shilin, I. V., Lunichkina, K. P. and Tsvetkova, Z. N., Zh. Neorg. Khim., 3, 2109 (1958)

59B Baldwin, W. H. Higgins, C. E. and Soldano, B. A., J. Phys. Chem., 63, 118 (1959)

59Ba Blake, C. A., Horner, D. E. and Schmidt, J. M., USAEC Report ORNL - 2259 (1959)

59C Carroll, W. R., Ph. D. Thesis, Univ. Missouri (1959)

59D Damiani, L. and Fattore, V., Energia Nucleare (Milan), 6, 793 (1959)

59H Hardy, C. J. and Scargill, D., J. Inorg. Nucl. Chem., 11, 128 (1959)

59Ha Healy, T. V., Kennedy, J. and Waind, G. M., J. Inorg. Nucl. Chem., 10, 137 (1959)

59Hb Healy, T. V. and Kennedy, J., J. Inorg. Nucl. Chem., 10, 308 (1959)

59Hc Hardy, C. J. and Greenfield, B. F., Unpublished work (1959) (quoted in 64H)

59I Irving, H. and Edgington, D. N., J. Inorg. Nucl. Chem., 10, 306 (1959)

59J Jury, S. H., and Whatley, M. E. USAEC Report CF-59-8-122 (1959)

59R Rozen, A. M. and Moiseenko, E. I., Zh. Neorg. Khim., 4, 1209 (1959)

59S Siddall, T. H., J. Amer. Chem. Soc., 81, 4176 (1959)

59Sa Sheka, Z. A. and Kriss, E. E., Zh. Neorg. Khim., 4, 2505 (1959)

59Sb Shevchenko, B. V., Solovkin, A. S., Shilin, I. V., Kirillov, L. M., Rodinov, A. V. and Balandina, V. V., Radiokhimiya, 1, 257 (1959)

59T Tuck, D. G., J. Chem. Soc., 218 (1959)

59V Voden, V. G. et al., Radiokhimiya, 1, 121 (1959)

59Va Vdovenko, V. M., Belov, L. M. and Chaikhorskii, A. A., Radiokhimiya, 1, 439 (1959)

60A Adamskii, N. M., Karpacheva, S. M., Mel'nikov, I. N. and Rozen, A. M., Radiokhimiya, 2, 13 (1960)

60Aa Adamskii, N. M., Karpacheva, S. M., Mel'nikov, I. N. and Rozen, A. M., Radiokhimiya, 2, 400 (1960)

60C Collopy, T. J. and Cavendish, J. H., J. Phys. Chem., 64, 1328 (1960)

60D Dyrssen, D. and Kuca, L., Acta Chem. Scand., 14, 1945 (1960)

60Da Dyrssen, D. and Liem, D. H., Acta Chem. Scand., 14, 1091 (1960)

60F Fomin, V. V. and Maiorova, E. P., Zh. Neorg. Khim., 5, 1100 (1960)

60Fa Fomin, V. V., Kartushova, R. E. and Maiorova, E. P., Zh. Neorg. Khim., 5, 1337 (1960)

60H Hardy, C. J., British Report AERE - R 3267 (1960)

60Ha Hesford, E. and McKay, H. A. C., J. Inorg. Nucl. Chem., 13, 156 (1960)

60Hb Higgins, C. E. and Baldwin, W. H., Anal. Chem., 32, 366 (1960)

60M Moiseenko, E. I. and Rozen, A. M., Radiokhimiya, 2, 274 (1960)

60Ma Marcus, Y., Israel Report IA 582 (1960)

60N Naito, K., Bull. Chem. Soc. Japan, 33, 363 (1960)

60P Pushlenkov, M. F., Komarov, E. V. and Shuvalov, O. N., Radiokhimiya, 2, 537 (1960)

60Pa Petrov, K. A., et al., Zh. Neorg. Khim., 5, 498 (1960)

60S Shevchenko, V. B., Shilin, I. V. and Zhdanov, Yu. F., Zh. Neorg. Khim., 5, 1366 (1960)

60Sa Shevchenko, B. V., Renard, E. V. and Solovkin, A. S., Zh. Neorg. Khim., 5, 2350 (1960)

60Sb Shevchenko, V. B., Shilin, I. V. and Zhdanov, Yu. F., Zh. Neorg. Khim., 5, 2832 (1960)

60Sc Shevchenko, V. B., Timoshev, V. G. and Volkova, A. A., Atomnaya Energia, 6, 426 (1960)

60Sd Solovkin, A. S., Thesis, Moscow, 1960 (quoted in 64S)

60Se Sato, T., J. Inorg. Nucl. Chem., 16, 156 (1960)

60Sf Shevchenko, V. B. Shmidt, V. S. and Mezhov, E. A., Zh. Neorg. Khim., 5, 1911 (1960)

60Sg Shevchenko, V. B. and Fedorov, I. A., Radiokhimiya, 2, 6 (1960)

60T Tsubomura, H. and Kliegman, J. M., J. Amer. Chem. Soc., 82, 1314 (1960)

60W Wilburn, N. P., USAEC Report HW - 66386 (1960)

61B Brauer, E. and Högfeldt, E., J. Inorg. Nucl. Chem., 23, 115 (1961)

61C Chaikhorskii, A. A., Vdovenko, V. M., Efimova, K. I. and Belov, L. M., Radiokhimiya, 3, 295 (1961)

61J Jezowska - Trzebiatowska, B. and Chmielowska, M., J. Inorg. Nucl. Chem., 20, 106 (1961)

61M Marcus, Y., J. Phys. Chem., 65, 1647 (1961)

61Ma Martin, B., Ockenden, O. W. and Foreman, J. K., J. Inorg. Nucl. Chem., 21, 96 (1961)

61Mb Moiseenko, E. I. and Rozen, A. M., Radiokhimiya, 3, 312 (1961)

61N Nomura, S. and Hara, R., Anal. Chim. Acta., 25, 212 (1961)

61O Olander, D. R., Donadieu, L. and Benedict, M., A. I. Ch. E. J., 7, 152 (1961) (quoted in 62D)

61S Sheka, Z. A. and Kriss, E. E., Zh. Neorg. Khim., 6, 1930 (1961)

61Sa Shevchenko, V. B., Shmidt, V. S. and Nenarokomov, E. A., Radiokhimiya, 3, 129 (1961)

61Sb Shevchenko, V. B., Fedorov, I. V. and Smelov, V. S., Radiokhimiya, 3, 256 (1961)

61Sc Shevchenko, V. B. and Fedorov, I. V., Radiokhimiya, 2, 6 (1960)

61U Umezawa, H. and Hara, R., Anal. Chim. Acta., 25, 360 (1961)

61Z Zemlyanukhin, V. I. and Savoskina, G. P., Radiokhimiya, 3, 411 (1961)

62A Arcand, G. M. and Carroll, W. R., J. Phys. Chem., 66, 1014 (1962)

62D Davis, W., Nucl. Sci. Eng., 14, 174 (1962)

62Da Davis, W., Nucl. Sci. Eng. , 14, 159 (1962)

62F Fomin, V. V., et al, Ekst. Teor. Primen. Appar. Sb. Statei., No. 2, p. 37, 1962

62Fa Fomin, V. V., Maiorova, E. P. and Kartushova, M. E., ibid, No. 1, p. 188

62M Morris, D. F. C. and Short, E. L., J. Chem. Soc., 2662 (1962)

62N Nikitina, G. P. and Pushlenkov, M. F., Radiokhimiya, 4, 137 (1962)

62P Prokhorova, N. P. and Brezhneva, N. E., Zh. Neorg. Khim., 7, 1846 (1962)

62Pa Pushlenkov, M. F., Komarov, E. V. and Shuvalov, O. N., Radiokhimiya, 4, 543 (1962)

62R Rozen, A. M. and Moiseenko, E. I., Ekst. Teor. Primen, Appar. Sb. Statei., 1962, No. 2, p. 236-56

62Ra Rozen, A. M., Khorkhorina, L. P., Karpacheva, S. M. and Agashkina, G. D., Radiokhimiya, **4**, 591 (1962)

62Rb Reznik, A. M., Rozen, A. M., Korovin, S. S. and Apraksin, I. A., Dokl. Adad. Nauk SSSR, 143, 1413 (1962)

62S Sheka, Z. A. and Kriss, E. E., Radiokhimiya, 4, 720 (1962)

62Sa Shevchenko, V. B., Smelov, V. S. and Strakhova, A. V., Ekst. Teor. Primen. Appar. Sb. Stat., 1962, No. 2, p. 179-89

62Sb Shevchenko, V. B., Smelov, V. S. and Strakhova, A. V., Zh. Neorg. Khim., 7, 1736 (1962)

62Sc Siekierski, S., J. Inorg. Nucl. Chem., 24, 205 (1962)

62Sd Smith, L. L., USAEC Report DP - 700 (1962)

62Y Yoshida, H., J. Inorg. Nucl. Chem., 24, 1257 (1962)

62Z Zemlyanukhin, V. I., Savoskina, G. P. and Pushlenkov, M. F., Radiokhimiya, 4, 570 (1962)

63A Apraksin, L. A., Korovin, S. S., Reznik, A. M. and Rozen, A. M., Zh. Neorg. Khim., 8, 237 (1963)

63B Baes, C. F., Jr., Nucl. Sci. Eng., 16, 405 (1963)

63K Komarov, E. V., Pushlenkov, M. F. and Shurenkova, M. E., Trudy Komis. Anal. Khim., Akad. Nauk. SSSR, Inst. Geokhim. i Anal. Khim., 14, 47 (1963)

63N Nishi, T. and Asano, M., Kogyo Kagaku Zasshi, 66, 1424 (1963)

63P Pushlenkov, M. F. and Shuvalov, O. N., Radiokhimiya, 5, 543 (1963)

63Pa Pushlenkov, M. F. and Shuvalov, O. N., Radiokhimiya, 5, 551 (1963

63R Rozen, A. M., Reznik, A. M., Korovin, S. S. and Metonidze, Z. A., Zh. Neorg. Khim., 8, 1003 (1963)

63Ra Reznik, A. M., Rozen, A. M., Korovin, S. S. and Apraksin, I. A., Radiokhimiya, 5, 49 (1963)

63Rb Rozen, A. M., Khorkhorina, L. P., Yurkin, V. G. and Novikova, N. M., Dokl. Akad. Nauk, SSSR, 153, 1387 (1963)

63S Starik, I. E. and Il'menkova, L. I., Radiokhimiya, 5, 236 (1963)

63Sa Shevchenko, V. B. and Renard, E. V., Zh. Neorg. Khim., 8, 516 (1963)

63Sb Smirnov - Averin, A. P., Kovalenko, G. S. and Krot, N. N., Zh. Neorg. Khim., 8, 2400 (1963)

63V Vdovenko, V. M., Lipovskii, A. A. and Nikitina, S. A., Radiokhimiya, 5, 139 (1963)

63W Whitney, D. C. and Diamond, R. M., J. Phys. Chem., 67, 209 (1963); Whitney, D. C., USAEC Report UCRL - 10505 (1962)

64A Apraksin, I. A., Korovin, S. S., Musorin, V. A. Reznik, A. M. and Rozen, A.M., Zh. Neorg. Khim., 9, 1295 (1964)

64Aa Aartsen, J. J. V. and Korvezee, A. E., Trans. Faraday Soc., 60, 510 (1964)

64Ab Afanas'ev, Yu. A., Radiokhimiya, 6, 499 (1964)

64B Burns, P. E. and Hanson, C., J. Appl. Chem., 14, 117 (1964)

64Ba Baram, I. I. and Laskorin, B. N., Zh. Neorg. Khim., 9, 738 (1964)

64D Dyrssen, D. and Liem, D. H., Acta Chem. Scand., 18, 224 (1964)

64E El - Guebeily, M. A., Hallaba, E. and Azzam, R.A.I., Proc. Int. Conf. Peaceful Uses At. Energ. Geneva, 10, 452 (1964)

64G Golub, A. M., Olevinskii, M. I. and Lutsenko, E. F., Ukr. Khim. Zh., 30, 1274 (1964)

64H Hahn, H. T. and Vander Wall, E. M., J. Inorg. Nucl. Chem., 26, 191 (1964)

64Ha Hertzog, D., French Report CEA - R - 2628 (1964)

64Hb Hardy, C. J., Fairhurst, D., McKay, H. A. C. and Willson, A. M., Trans. Faraday Soc., 60, 1626 (1964)

64K Kubala, J. and Mazonska, D., Roczniki Chem., 38, 527 (1964)

64L Lipovskii, A. A. and Yakovleva, N. E., Zh. Neorg. Khim., 9, 767 (1964)

64N Nikolaev, A. V. and Afanas'ev, Yu. A., Dokl. Akad. Nauk. SSSR, 155, 374 (1964)

64Na Nikolaev, A. V., Torgov, V. G., Mikhailov, V. A. and Kotlyarevskii, I. L., Dokl. Akad. Nauk SSSR, 156, 616 (1964)

64Nb Nikolaev, A. V., Avanas'ev, Yu. A., Ryabinin, A. L. and Koroleva, T. I., Dokl. Akad. Nauk SSSR, 159, 851 (1964)

64Nc Nishimura, S., Moriyama, J. and Kushima, I., Trans. Japan Inst. Metals, 5, 79 (1964)

64P Pushlenkov, M. F. and Komarov, E. V., Radiokhimiya, 6, 426 (1964)

64R Rozen, A. M. and Nikolotova, Z. I., Zh. Neorg. Khim., 9, 1725 (1964)

64S Solovkin, A. S., Zh. Neorg. Khim., 9, 746 (1964)

64Sa Sato, T., J. Inorg. Nucl. Chem., 26, 311 (1964)

64Sb Specker, H., Werding, G. and Shiewe, G., Z. Anal. Chem., 206, 161 (1964)

64Sc Siekierski, S., Nukleonika, 9, 601 (1964)

64Sd Shuvalov, O. N., Radiokhimiya, 7, 265 (1964)

64Se Shevchenko, F. D., Dopovidi Akad. Nauk Ukr. RSR, 1964 (4), p. 511

64V Vdovenko, V. M. and Bulyamitsa, L. S., Radiokhimiya, 6, 399 (1964)

64Z Zemlyanukhin,V. I., Savoskina, G. P. and Pushlenkov, M. F., Radiokhimiya, 6, 714 (1964)

65A Afanas'ev, Yu. A., Ryabinin, A. I. and Koroleva, T. I., Zh. Fiz. Khim., 39, 2296 (1965)

65B Belyaev, A. V. and Ptitsyn, B. V., Izv. Sib. Otd. Akad. Nauk SSSR. Ser. Khim. Nauk., 1965 (3) p. 144

65D Dyachkova, R. A., Spitsyn, V. I. and Khlebnikov, V. P., Radiokhimiya, 7, 262 (1965)

65Da Dyrssen, D. and Petkovic, D. J., J. Inorg. Nucl. Chem., 27, 1381 (1965)

65F Fletcher, J. M., Lyon, C. E. and Wain, A. G., J. Inorg. Nucl. Chem., 27, 1841 (1965)

65G Golub, A. M., Olevinskii, M. I. and Lutsenko, E. F., Ukr. Khim. Zh., 31, 12 (1965)

65K Kozlova, M. D. and Levin, V. I., Radiokhimiya, 7, 534 (1965)

65Ka Korovin, S. S., Lebedeva, E. N., Dedich, K., Reznik, A. M. and Rozen, A. M., Zh. Neorg. Khim., 10, 518 (1965)

65Kb Kuzin, I. A. and Chuchalin, L, K,, Zh. Prikl. Khim., 38, 2422 (1965)

65L Levin, V. I. and Kozlova, M. D., Radiokhimiya, 7, 437 (1965)

65La Lipovskii, A. A., Nikitina, S. A. and Yakovleva, N. E., Radiokhimiya, 7, 563 (1965)

65N Nikolaev, A. V., Afanas'ev, Yu. A. and Durasov, V. B., Dokl. Akad. Nauk SSSR, 162, 1317 (1965)

65Na Nishimura, S., Tokura, J. and Kondo, Y., Mem. Fac. Eng. Kyoto Univ., 27, 202 (1965)

65R Rozen, A. M., Nikolotova, Z. I., Petrov, K. A., Skotnikov, A. S. and Teterin, E. G., Radiokhimiya, 7, 517 (1965)

65S Samodelov, A. P., Z. Neorg. Khim., 10, 2180 (1965)

65Sa Starik, I. E. and Ampelogova, N. I., Radiokhimiya, 7, 658 (1965)

65Sb Shvedov, V. P. and Orlov, Yu. F., Zh. Neorg. Khim., 10, 2774 (1965)

65W Woodhead, J. L. and McKay, H. A. C., J. Inorg. Nucl. Chem., 27, 2247 (1965)

66E Estrada, J., Diaz Nogueira, E. and Lopez Perez, B., An. Real Soc. Espana Fis. Quim. (Madrid), Ser. B62 (3)

66K Kholebnikov, V. P., D'yachkova, R. A. and Spitsyn, V. I., Radiokhimiya, 8, 125 (1966)

66Ka Korovin, S. S., Yurkin, V. G. and Mironenko, A. P., Zh. Neorg. Khim., 11, 1910 (1966)

66Kb Korovin, S. S., Kol'tsov, Yu. I., Reznik, A. M. and Apraksin, I. A., Zh. Neorg. Khim., 11, 180 (1966)

66Kc Kol'tsov, Yu. I., Korovin, S. S. and Reznik, A. M., Zh. Neorg. Khim., 11, 2308 (1966)

66L Liem, D. H. and Dyrssen, D., Acta Chem. Scand., 20, 272 (1966)

66La Levin, V. I. and Kozlova, M. D., Radiokhimiya, 8, 533 (1966)

66R Reznik, A. M., Potapov, G. G., Korovin, S. S. and Apraksin, I. A., Zh. Neorg. Khim., 11, 1894 (1966)

66S Szabo, E. and Szabon, S., Acta Chem. Acad. Sci. Hungaricae, 48, 299 (1966)

66W Wang, S. M. and Li, N. C., J. Inorg. Nucl. Chem., 28, 1091 (1966)

66Z Zangen, M., J. Inorg. Nucl. Chem., 28, 1693 (1966)

67B Biddle, P., Coe, A., McKay, H. A. C., Miles, J. H. and Waterman, M. J., J. Inorg. Nucl. Chem., 29, 2615 (1967)

67D Davis, W. Jr. and Mrochek, J., in "Solvent Extraction Chemistry" eds. Dyrssen, D., Liljenzin, J. O. and Rydberg, J., North-Holland, Amsterdam, 1967, p. 283

67G Golub, A. M. and Borshch, A. N., Zh. Neorg. Khim., 12, 522 (1967)

67J Johnson, J. R., Christian, S. D. and Affsprung, H. E., J. Chem. Soc., 1924 (1967)

67K Kuca, L., Coll. Czech. Chem. Comm., 32, 288 (1967)

67Ka Kuca, L., Coll. Czech. Chem. Comm., 32, 720 (1967)

67Kb Korovin, S. S., Lebedeva, E. N., Rozen, A. M. and Torubarov, D. I., Zh. Neorg. Khim., 12, 537 (1967)

67Kc Kettrup, A. and Specker, H., Z. Anal. Chem., 230, 241 (1967)

67Kd Krasovec, F. and Klofutar, C. in "Solvent Extraction Chemistry", eds. Dyrssen, D., Liljenzin, J. O. and Rydberg, J., North-Holland, Amsterdam 1967, p. 509

67L Liem, D. H., in "Solvent Extraction Chemistry" eds. Dyrssen, D., Liljenzin, J. O. and Rydberg, J., North-Holland, Amsterdam, 1967, p. 264

67M Mikhailov, V. A., Grigoreva, E. F. and Kharchenko, S. K., Radiokhimiya, 9, 149 (1967)

67N Nikolaev, A. V., Afanas'ev, Yu. A. and Starosin, A. D., Dokl. Akad. Nauk. SSSR, 176, 110 (1967)

67P Petkovic, Dj. M. in "Solvent Extraction Chemistry" eds. Dyrssen, D., Liljenzin, J. O. and Rydberg, J., North-Holland, Amsterdam 1967, p. 305

67R Rozen, A. M., Khorkhorina, L. P. and Yurkin, V. G., Zh. Neorg. Khim., 12, 244 (1967)

67Ra Rozen, A. M. and Mikhailichenko, A. I., Zh. Neorg. Khim., 12, 741 (1967)

67S Siddall, T. H., III and Stewart, W. E., Inorg. Nucl. Chem. Letters, 3, 279 (1967)

67V Vorob'ev, S. P., Davydov, I. V. and Shilin, J. V., Zh. Neorg. Khim., 12, 2665 (1967)

67Va Vdovenko, V. M. and Lipovskii, A. A. in "Solvent Extraction Chemistry" eds. Dyrssen, D., Liljenzin J. O. and Rydberg, J., North-Holland, Amsterdam, 1967, p. 307

67Z Zelikman, A. N. and Dregan, L., Zh. Neorg. Khim., 12, 261 (1967)

68A Asano, M., Tech. Rep. Eng. Res. Inst. Kyoto Univ. No. 138 (1968)

68Aa Alian, A. and Zaborenko, K. B., Z. Anal. Chem., 238, 267 (1968)

68C Chifu, E., Gazz. Chim. Ital., 98, 1213 (1968)

68G Gorbenko, F. P., Kuchlina, E. D. and Olevinskii, M. I., Zh. Anal. Khim., 23, 1301 (1968)

68H Hallaba, E. and Azzam, R., Z. Phys. Chem., 237, 177 (1968)

68K Kuznetsova, N. H., Zh. Anal. Khim., 23, 1485 (1968)

68Ka Krasovec, F. and Klofutar, C., J. Inorg. Nucl. Chem., 30, 1640 (1968)

68L Liem, D. H., Acta Chem., Scand., 22, 773 (1968)

68La Liem, D. H., Acta Chem. Scand., 22, 753 (1968)

68Lb Lapitskaya, E. V. and Gorbenko, F. P., Radiokhimiya, 10, 90 (1968)

68Lc Lapitskaya, E. V., Kuchkina, E. D. and Gorbenko, F. P., Zh. Neorg. Khim., 13, 2774 (1968)

68Ld Liem, D. H., in "Progress in Coordination Chemistry" (M. Cais Ed.), Elsevier, Amsterdam 1968, p. 740

68M Mikhailov, V. A., Ponomareva, L. I. and Nazin, A. G., Zh. Fiz. Khim., 42, 318 (1968)

68Ma Mitamura, R., Tokura, I., Nishimura, S., Kondo, Y. and Li, N. C., J. Inorg. Nucl. Chem., 30, 1019 (1968)

68N Nishimura, S., Ke, C. H. and Li, N. C., J. Amer. Chem. Soc., 90, 234 (1968)

68Na Nishimura, S., Mitamura, R., Kondo, Y. and Li, N. C., J. Inorg. Nucl. Chem., 30, 3033 (1968)

68O Ochsenfeld, W., Schmieder, H. and Baumgärtel, G., German report KFK - 671 (1968)

68P Petkovic, Dj. M., J. Inorg. Nucl. Chem., 30, 603 (1968)

68R Roland, G., and Duyckaerts, G., Spectrochim. Acta, 24A, 529 (1968)

68S Startsev, V. N., Sannikov, Yu. I., Ben'yash, G. N. and Krylov, E. I., Zh. Neorg. Khim., 13, 1122 (1968)

68Sa Startsev, V. N., Sannikov, Yu. I. and Stroganov, S. S., Zh. Neorg. Khim., 13, 1608 (1968)

68Sb Shikheeva, L. V., Zh. Neorg. Khim., 13, 3323 (1968)

69B Bucher, J. J. and Diamond, R. M., J. Phys. Chem., 73, 675 (1969)

69H Hanson, C. and Patel, A. N., J. Appl. Chem., 19, 20 (1969)

69M Meier, H., et al. Mikrochim. Acta, 557 (1969)

69N Nakamura, K., J. Inorg. Nucl. Chem., 31, 455 (1969)

69S Shehata, M. K. K. and Aten, Jr. A. H. W., J. Inorg. Nucl. Chem., 31, 239 (1969)

69Y Yagodin, G. A. and Tarasov, V. V., Radiokhimiya, 11, 148 (1969)

METAL ION, M	REACTION	LOG K	TEMP.	CONDITIONS, AQ.	CONDITIONS, ORG.	REF.
Extractant: **Triisobutyl** Phosphate, (TiBP), S						
Ligand: NO_3^- , A^-						
	$H^+ + A^- + \overline{S} = \overline{HA \cdot S}$	ΔG 1.44	25	HA 1.0	S 0.05 in n-dodecane	59S
		ΔH-0.7				
		ΔS-7.2				
UO_2^{++}	$M^{++} + 2A^- + 2\overline{S} = \overline{MA_2 \cdot 2S}$	ΔG-1.74 ΔH-6.6 ΔS-16.2	25	HA 1.0	S 0.05 in n-dodecane	59S
		$-\Delta G = -16.24T + 6.579 \cdot 10^3$				
Ligand: SCN^- , A^-						
	$\overline{A} + \overline{S} = \overline{A \cdot S}$	1.2	25	——	S $1.5 \cdot 10^{-2}$ + A 0.1-0.2 in benzene	69D
		1.12	35			
		1.00	50			
		ΔH-2.9	25-50			

References

59S Sidall, T. H., J. Amer. Chem. Soc., 81, 4176 (1959)

69D Dahl, R., Klaboe, P. and Gramstad, T., Spectrochim. Acta, 25A, 207 (1969)

METAL ION, M	REACTION	LOG K	TEMP.	CONDITIONS, AQ.	CONDITIONS, ORG.	REF.
Extractant: Tri-Octyl Phosphate, TOP, S						
Ligand: NO_3^- , A^-						
	$H^+ + A^- + \overline{S} = \overline{HA \cdot S}$	-0.62	25	HA 0.1-2.0	S 1.0 in benzene	65S
		-0.72	25	HA 0.5	S 0.5 in CCl_4	64R
UO_2^{++}	$M^{++} + 2A^- + 2\overline{S} = \overline{MA_2 \cdot 2S}$	1.6 to 1.8	25	HA 0.05	S 0.05 in CCl_4	64R
Ligand: ClO_4^- , A^-						
	$\overline{S} + \overline{(H)DpCl\text{-}PhP} = \overline{S \cdot (H)DpCl\text{-}PhP}$	3.38	r.t.	HA 0.1	$S \gg 3.5 \times 10^{-4} > (H)DpCl\text{-}PhP$ in CCl_4	68K

References

64R Rozen, A. M. and Nikolotova, Z. I., Zh. Neorg. Khim., 9, 1725 (1964)

65S Shvedov, V. P. and Orlov, Yu. F., Zh. Neorg. Khim., 10, 2774 (1965)

68K Krasovec, F. and Klofutar, C., J. Inorg. Nucl. Chem., 30, 1640 (1968)

METAL ION, M	REACTION	LOG K	TEMP.	CONDITIONS, AQ.	CONDITIONS, ORG.	REF.
	Extractant Class: Phosphoric and Pyrophosphoric acids					
Extractant: Dibutyl phosphoric acid (HDBP), HS						
Distribution Constants						
	$HS = \overline{HS}$	0.34	25	$(Na^+,H^+)ClO_4^-$ 0.1	TBP<3.4 in $CHCl_3$	57Da
		1.36		$(Na^+,H^+)ClO_4^-$ 0.1	HS<1 in MiBK	
		1.49		$(Na^+,H^+)ClO_4^-$ 1	HS<1 in MiBK	
		1.87		$HClO_4$ 1	HS<3.10^{-5} TBP<3.4 in $CHCl_3$	
		-1.96		HNO_3 1	HS<0.05 in kerosene	59Ha
		-0.91		HNO_3 1	HS<0.05 in CCl_4	
		-0.42			HS<0.05 in benzene	
		-0.14		HNO_3 1	HS<0.05 in nitrobenzene	
		0.28		$HClO_4$ 1	HS<0.05 in $CHCl_3$	
		-0.14		$HClO_4$ 0.1	HS<0.61 in butylether	59Da
		0.52		HNO_3 0.1	HS<0.01 in isopropylether	60Da
		2.21			HS<0.01 in MiBC	
		1.97	25		HS<0.01 in TBP in CCl_4	

METAL ION, M	REACTION	LOG K	TEMP.	CONDITIONS, AQ.	CONDITIONS, ORG.	REF.
Distribution Constants (cont.)						
	$HS = \overline{HS}$	0.24	25	HNO_3 0.1	HS<0.01 in $CHCl_3$	60Da
		-1.44		HNO_3 0.1	HS<0.01 in CCl_4	
		-2.34		HNO_3 0.1	HS<0.01 in n-hexane	
		-0.7		$HClO_4$ 0.1	HS<0.6 in CCl_4	60Dc
		-0.70		HNO_3 1	HS<0.5 in toluene	61G
		-0.75	?	$HClO_4$ 1.6	HS<0.8 in CCl_4	63A
		-0.11	?	$HClO_3$ 3	HS<0.8 in CCl_4	
		0.45	?	$HClO_4$ 6	HS<0.8 in CCl_4	
		0.62	?	HCl 2.5	HS<0.8 in CCl_4	
		0.61	25	HNO_3 4, $NaNO_3$ 2	HS<0.01 in benzene	63D
		-2.33	?	$(Na^+,H^+)ClO_4^-$ 0.1	HS<10^{-4} in n-octane	65U,67Kc
		-2.13	25	H_2SO_4 0.1	HS<0.1 in benzene	66D,67L
		-1.36	25	H_2SO_4 0.1	HS<0.1 in CCl_4	66D,67L, 68L
		0.9-1.36	?	H_2O, pH=2-3	HS<0.3 in n-butanol	66K

METAL ION, M	REACTION	LOG K	TEMP.	CONDITIONS, AQ.	CONDITIONS, ORG.	REF.
Distribution Constants (cont.)						
	$HS = \overline{HS}$	-1.32	?	HNO_3 0.1, $NaNO_3$ 0.9	HS<0.06 in CCl_4	67K
		-1.29	?	HNO_3 1.0		
		1.11	?	HNO_3 6		
		-1.21	?	$HClO_4$ 0.1, $NaClO_4$ 0.9	HS<0.06 in CCl_4	
		0.4	?	HNO_3 0.1, $NaNO_3$ 5.9	HS<0.06 in CCl_4	
		0.12	?	HNO_3 , $NaNO_3$ 5	HS<0.06 in CCl_4	
	$H_2S_2 = \overline{H_2S_2}$	4.1	25	$HClO_4$ 0.1	HS<5.10^{-3} in $CHCl_3$	57Da
		2.8			in hexane	
		0.8		HNO_3 1	HS<0.05 in kerosene	59Ha
		2.4			in CCl_4	
		2.9			in benzene	
		2.2			in $CHCl_3$	
		3.8		$HClO_4$ 0.1		
		2.6		HNO_3 1	HS<0.5 in toluene	61G
		3.04	?	$HClO_4$ 1.6	HS<0.8 in CCl_4	63A

METAL ION, M	REACTION	LOG K	TEMP.	CONDITIONS, AQ.	CONDITIONS, ORG.	REF.
Distribution Constants (cont.)						
	$H_2S_2 = \overline{H_2S_2}$	2.39	?	$HClO_4$ 3	HS<0.8 in CCl_4	63A
		2.88	?	$HClO_4$ 6		
		2.75	?	HCl 2.5		
Dissociation						
	$HS = H^+ + S^-$	-1.7 to -1.14	?	H_2O pH= 2.3	HS 0.3 in n-butanol	66K
		-1.72	25	H_2O,HS 0.23	———	43K
		-1.13	?	(Na^+,H^+) NO_3^- 1.0	HS 10^{-3} in CCl_4	67K
		-1.20	?	(Na^+,H^+) NO_3^- 1.0	HS<$3 \cdot 10^{-5}$ in CCl_4	
		-3.00	?	$(Na^+ H^+)$ NO_3^- 6.0	HS<10^{-3} in CCl_4	
		-1.25	?	HCl 2.5	HS<0.8 in CCl_4	63A
		-1.17	25	μ=1 (Na^+,H^+) ClO_4^-	———	57Da
		-1.14	?	$HClO_4$ 1.6	HS<0.8 in CCl_4	63A
		-1.33	?	$HClO_4$ 3.0	HS<0.8 in CCl_4	
		-1.85	?	$HClO_4$ 6.0	HS<0.8 in CCl_4	

E

METAL ION, M	REACTION	LOG K	TEMP.	CONDITIONS, AQ.	CONDITIONS, ORG.	REF.
Dissociation (contd.)						
	$HS = H^+ + S^-$	-1.00	20	$(Na^+,H^+)\ ClO_4^-$ 0.1	HS $3\cdot10^{-3}$ in octane	65U
		-1.04	?	$(Na^+,H^+)\ ClO_4^-$ 1.0	HS 10^{-3} in CCl_4	67K
		-1.18	?	$(Na^+,H^+)\ ClO_4^-$ 1.0	HS $3\cdot10^{-5}$ in CCl_4	
	$\overline{HS} = \overline{H^+} + \overline{S^-}$	-3.75	?	———	HS 0.05 in 98% ethanol	65D
Dimerization						
	$2HS = H_2S_2$	4.7	?	HNO_3 0.1, $NaNO_3$ 5.9	HS<0.03 in CCl_4	67K
		4.3	?	HNO_3 1.0, $NaNO_3$ 5.0	HS<0.03 in CCl_4	
		1.50	?	HNO_3 6.0	HS<0.03 in CCl_4	
		1.11	25	sat. solution of HS in water		57Da
		2.36	25	H_2SO_4 0.1	HS<0.5 in CCl_4	66D,67L, 68L
		2.33	?	HCl 2.5	HS<0.8 in CCl_4	63A
		0.70	?	$HClO_4$ 1.6	HS<0.8 in CCl_4	
		2.28	?	$HClO_4$ 3.0	HS<0.8 in CCl_4	
		2.66	?	$HClO_4$ 6.0	HS<0.8 in CCl_4	

METAL ION, M	REACTION	LOG K	TEMP.	CONDITIONS, AQ.	CONDITIONS, ORG.	REF.
Dimerization (cont.)						
	$\overline{2HS} = \overline{H_2S_2}$	4.48	25	$HClO_4$ 0.1	HS<1.0 in $CHCl_3$	57Da
		4.36	25	$HClO_4$ 1.0	HS<$5.5 \cdot 10^{-3}$ in $CHCl_3$	59Ha
		4.21	25	$HClO_4$ 0.1	HS<$5 \cdot 10^{-2}$ in $CHCl_3$	
		4.61	25	HNO_3 0.1	HS<0.01 in $CHCl_3$	60Da
		4.84			HS in CCl_4	58D
		5.33	25	HNO_3 1.0	HS<0.05 in CCl_4	59Ha
		6.49	25	HNO_3 0.1	HS<0.5 in CCl_4	60Da
		4.85	?	HCl 2.5	HS<0.8 in CCl_4	63A
		5.25	?	$HClO_4$ 1.6	HS<0.8 in CCl_4	
		4.77	?	$HClO_4$ 3.0	HS<0.8 in CCl_4	
		4.63	?	$HClO_4$ 6.0	HS<0.8 in CCl_4	
		6.27	25	H_2SO_4 0.1	HS<0.5 in CCl_4	66D
		6.37	25	H_2SO_4 0.1	HS<0.5 in CCl_4	67L,68L/ 66D
		6.19	?	$HClO_4$ 0.1, $NaClO_4$ 0.9	HS<0.03 in CCl_4	67K
		6.01	?	$HClO_4$ 1.0	HS<0.03 in CCl_4	

METAL ION, M	REACTION	LOG K	TEMP.	CONDITIONS, AQ.	CONDITIONS, ORG.	REF.
Dimerization (cont.)						
	$\overline{2HS} = \overline{H_2S_2}$	6.28	?	HNO_3 0.1, $NaNO_3$ 0.9	HS<0.03 in CCl_4	67K
		6.1	?	HNO_3 1.0	HS<0.03 in CCl_4	
		5.0	?	HNO_3 1.0, $NaNO_3$ 5.0	HS<0.03 in CCl_4	
		5.6	?	HNO_3 6.0	HS<0.03 in CCl_4	
		3.6	25	$HClO_4$ 0.1	$2\cdot10^{-5}$<< HS<0.61 in dibutyl ether	59Da
		6.87	25	HNO_3 0.1	HS<0.5 in n-hexane	60Da
		6.66-6.52	25	H_2SO_4 0.1	HS<0.1 in hexane	66D,67L, 68L
		< -1.0	25	HNO_3 0.1	0.45 in MiBK	60Da
		2.29	25	HNO_3 0.1	HS<0.44 isopropyl ether	
		6.86	20	$(Na^+,H^+)ClO_4^-$ 0.1 pH = 1.05	HS<$3\cdot10^{-3}$ in n-octane	65U,67Kc
		1.19	25	$HClO_4$ 0.1	HS<1.0 in MiBK	57Da
		5.63		HNO_3 1.0	HS<0.05 in kerosene	64H/59Ha
		5.78	25	HNO_3 1.0	HS<0.05 in kerosene	59Ha
		4.88	25	HNO_3 1.0	HS<0.05 in benzene	
		3.55	25	HNO_3 1.0	HS<0.02 in nitrobenzene	

METAL ION, M	REACTION	LOG K	TEMP.	CONDITIONS, AQ.	CONDITIONS, ORG.	REF.
Dimerization (cont.)						
	$\overline{2HS} = \overline{H_2S_2}$	5.09	25	HNO_3 1.0	HS<0.5 in toluene	61G
		-0.12 (?)	25	HNO_3 0.1	HS<0.41 in TBP	60Da
	$4\overline{HS} = \overline{H_4S_4}$	14.62	25	H_2SO_4 0.1	HS>0.1 in hexane	67L,68L
Reactions with diluents						
	$\overline{H_2S_2} + \overline{CHCl_3} = \overline{H_2S_2 \cdot CHCl_3}$	1.80	25	HNO_3 0.1	HS<0.5 in $CHCl_3$	64D/57Da, 60Da
	$\overline{HS} + \overline{CHCl_3} = \overline{HS \cdot CHCl_3}$	1.48	25	HNO_3 0.1	HS<0.5 in $CHCl_3$	
		0.53	25	HNO_3 0.1	HS<0.6, $CHCl_3$ 12.53 sat. with water	60Da
	$\overline{HS} + \overline{CCl_4} = \overline{HS \cdot CCl_4}$	K ∿ 0	25	HS<0.5	CCl_4 10.37 sat. with water	
		-0.17	25	HNO_3 0.1	HS<0.5 in CCl_4	64D/57Da. 60Da
	$\overline{H_2S_2} + \overline{CCl_4} = \overline{H_2S_2 \cdot CCl_4}$	0.39	25	HNO_3 0.1	HS<0.5 in CCl_4	
	$\overline{HS} + \overline{n\text{-}C_6H_{14}} = \overline{HS \cdot n\text{-}C_6H_{14}}$	K ∿ 0	?	HS<0.5	hexane 7.66 sat. with water	60Da
		0.0	25	HNO_3 0.1	HS<0.5 in hexane	64D/57Da, 60Da

METAL ION, M	REACTION	LOG K	TEMP.	CONDITIONS, AQ.	CONDITIONS, ORG.	REF.
Reactions with diluents (cont.)						
	$\overline{H_2S_2} + \overline{nC_6H_{14}} = \overline{H_2S_2 \cdot nC_6H_{14}}$	0.0	25	HNO_3 0.1	HS<0.5 in hexane	64D/57Da, 60Da
	$\overline{HS} + \overline{C_6H_6} = \overline{HS \cdot C_6H_6}$	-0.12	25	HNO_3 0.1	HS ? C_6H_6 11.25 sat. with water	60Da
	$\overline{HS} + \overline{MiBC} = \overline{HS \cdot MiBC}$	2.74	25	HNO_3 0.1	HS<0.5, MiBC 7.39 sat. with water	
	$\overline{HS} + \overline{MiBK} = \overline{HS \cdot MiBK}$	0.75	25	μ = 0.1 and 1.0 $(H^+,Na^+)ClO_4$	HS $1.3 \cdot 10^{-5}$ in MiBK	57Da
		1.86	25	HNO_3 0.1	MiBK 7.82 sat. with water	60Da
	$\overline{HS} + \overline{(CH_3)_2CHOCH(CH_3)_2} = \overline{HS \cdot (CH_3)_2CHOCH(CH)_3}$	1.07	25	HNO_3 0.1	HS<0.5 isopropylether sat. with water	
	$\overline{HS} + \overline{C_6H_5NO_2} = \overline{HS \cdot C_6H_5NO_2}$	0.25	25	HNO 0.1	HS ? in nitrobenzene sat. with water	
Various reactions in org. phase with other org. phosphates						
	$\overline{HS} + \overline{TBP} = \overline{HS \cdot TBP}$	1.60	25	$HClO_4$ 1.0	HS $3 \cdot 10^{-5}$. TBP<0.4 in $CHCl_3$	57Da
		2.83	25	HNO_3 1.0	HS 10^{-5}, TBP 0.3 in kerosene	59Ha
		2.84	25	HNO_3 1.0	TBP<3.43, HS<10^{-4} in CCl_4	60Da
		4.20		HNO_3 1.0	HS 10^{-5}, TBP 0.69 in kerosene	64H/59H

METAL ION, M	REACTION	LOG K	TEMP.	CONDITIONS, AQ.	CONDITIONS, ORG.	REF.
Various reactions in org. phase with other org. phosphates (cont.)						
	$\overline{HS} + \overline{TBP} = \overline{HS \cdot TBP}$	3.11		HNO_3 1.0	HS 10^{-5} , TBP 0.35 in kerosene	64H/59H
		3.77	25	HNO_3 0.1	HS<0.5 in hexane	64D/57Da, 60Da
		2.87	25	HNO_3 0.1	HS<0.5 in CCl_4	
		1.19	25	HNO_3 0.1	HS<0.5 in $CHCl_3$	
		2.90	25	H_2SO_4 0.1	HS + TBP in hexane	66D,67L
		2.63	25	H_2SO_4 0.1	HS + TBP in CCl_4	
		2.66	?	HNO_3 0.1, $NaNO_3$ 0.9	HS<0.003 + TBP<0.5 in CCl_4	67Kb
	$\overline{HS} + \overline{2TBP} = \overline{HS \cdot (TBP)_2}$	3.30	25	H_2SO_4 0.1	HS + TBP in hexane	66D,67L
	$\overline{H_2S_2} + \overline{TBP} = \overline{H_2S_2 \cdot TBP}$	1.08	25	HNO_3 0.1	HS<0.5 in hexane	64D/57Da, 60Da
		-0.55 (?)	25	HNO_3 0.1	HS 0.5 in $CHCl_3$	
		-0.75	25	HNO_3 0.1	HS 0.5 in CCl_4	
		1.23	25	H_2SO_4 0.1	HS + TBP in hexane	67L
		0.22	25	H_2SO_4 0.1	HS + TBP in CCl_4	67L

METAL ION, M	REACTION	LOG K	TEMP.	CONDITIONS, AQ.	CONDITIONS, ORG.	REF.
Various reactions in org. phase with other org. phosphates (cont.)						
	$\overline{HS} + 0.75\ \overline{TH'PO} = \overline{HS \cdot 0.75\ TH'PO}$	3.4	?	$(H^+, Na^+)\ NO_3^-$ 1.0	HS<0.003 + TH'PO<0.5 in CCl_4	67Kb
	$\overline{HS} + 0.5\ \overline{TH'PO} = \overline{HS \cdot 0.5\ TH'PO}$	2.5	?	$(H^+, Na^+)\ NO_3^-$ 6.0	HS<0.003, TH'PO<0.5 in CCl_4	
		2.9	?	HNO_3 0.1 $NaNO_3$ 5.9		
	$\overline{HS} + \overline{TO'PO} = \overline{HS \cdot TO'PO}$	4.82	25	H_2SO_4 0.1	HS + TO'PO in hexane	66D,67L
		4.29	25	H_2SO_4 0.1	HS + TO'PO in CCl_4	
Ligand: H_2O						
	$\overline{(HS)_2} + H_2O = \overline{(HS)_2 \cdot H_2O}$	-0.4	19-25	H_2O or HNO_3 1.0 or H_2SO_4	HS 0.1-0.01 in CCl_4	67Kd
Ligand: HDBP, HS						
Y^{+3}	$M^{+3} + S^- = MS^{+2}$	2.91	20	HNO_3<0.5	HS<0.1 in benzene	61S
La^{+3}	$MS_{3\ solid} = MS_{3\ aq.}$ or:	ΔH 3.9				
	$MS_{3\ solid} + mH_2O = M(H_2O)_x^{+3} + MS(H_2O)_{x-2}^{+2}$	ΔG 14.35	25	solubility in MS_3 in H_2O		64S
	$+ 2S(H_2O)_y$	ΔS 61.5				
	$M^{+3} + S^- = MS^{+2}$	1.78	25	HNO_3<0.3		64Sa
		2.0	25	HS aq. $2.0 \cdot 10^{-5}$ (MS_3 sat. solution in H_2O) $\mu \to 0$		65S

METAL ION, M	REACTION	LOG K	TEMP.	CONDITIONS, AQ.	CONDITIONS, ORG.	REF.
Ligand: HDBP , HS (cont.)						
La^{3+}	$M^{+3} + S^{-} = MS^{+2}$	1.80	?	$\mu \to 0$, pH<1.3 HS aq<0.08		65Sa
		1.42	25	$NaNO_3$ 0.01 sat. aq. solution of MS_3		65Sb
		1.10	25	$NaNO_3$ 1.0 sat. aq. solution of MS_3		
	$M^{+3} + 2S^{-} = MS_2^{+}$	∿ 3.5	25	$\mu \to 0$ pH<1.3 HS aq.<0.08		65Sa,66S
		1.69	25	$NaNO_3$ 1.0, sat. aq. solution of MS_3		65Sb
		2.56	25	$NaNO_3$ 0.01 sat. aq. solution of MS_3		
	$M^{+3} + 3S^{-} = MS_3$	4.62	?	$\mu \to 0$ pH<1.3 HS aq.<0.08		65Sa
		3.86	25	$NaNO_3$ 0.01, sat. aq. solution of MS_3		65Sb
		2.78	25	$NaNO_3$ 1.0, sat. aq. solution of MS_3		
Ce^{+3}	$M^{+3} + S^{-} = MS^{+2}$	2.5	20	HNO_3<0.5	HS<0.1 in benzene	61S
Pr^{+3}	$M^{+3} + S^{-} = MS^{+2}$	2.17	25	HS aq. $2.0 \cdot 10^{-5}$ (MS_3 sat. sol. in aq.) $\mu \to 0$		65S

METAL ION, M	REACTION	LOG K	TEMP.	CONDITIONS, AQ.	CONDITIONS, ORG.	REF.
Ligand: HDBP HS (cont.)						
Nd^{+3}	$MS_3(s) = MS_3\ aq.$ or: $MS_3(s) + mH_2O = M(H_2O)_x^{+3} + MS(H_2O)_{x-2}^{+2} + 2S(H_2O)_y^-$	ΔH 5.7 ΔG 16.79 ΔS 75.5	25	solubility of MS_3 in H_2O	——	64S
	$M^{+3} + S^- = MS^{+2}$	2.29	25	HS aq. $2.0\cdot10^{-5}$ (MS_3 sat. sol. in aq.) $\mu \to 0$		65S
	$\overline{MS_3} + 1.5\overline{(HS)_2} = \overline{M(HS_2)_3}$	-0.41	25	——	HS<1 in benzene	63K
Sm^{+3}		2.48	25			
Eu^{+3}		2.61	25			
Gd^{+3}	$MS_3(s) = MS_3\ aq.$ or: $MS_3(s) + mH_2O = M(H_2O)_x^{+3} + MS(H_2O)_{x-2}^{+2} + 2S(H_2O)_y^-$	ΔH -5.30 ΔG 19.38 ΔS -83	25	solubility of MS_3 in H_2O	-	64S
	$M^{+3} + S^- = MS^{+2}$	2.73	25	HS aq. $2.0\cdot10^{-5}$ (MS_3 sat. sol. in aq.) $\mu \to 0$		65S
Tb^{+3}		2.78	25			
Ho^{+3}		3.00	25			
Er^{+3}		3.11	25			
Yb^{+3}	$\overline{MS_3} + 1.5\ \overline{(HS)_2} = \overline{M(HS_2)_3}$	-0.95	20	-	HS<1.0 in CCl_4	61Ka,62S, 63K
		-1.14	25	-	HS<1.0 in benzene	

METAL ION, M	REACTION	LOG K	TEMP.	CONDITIONS, AQ.	CONDITIONS, ORG.	REF.
Ligand: HDBP , HS (cont.)						
Yb^{+3}	$MS_3(s) = MS_3\ aq.$ or: $MS_3(s) + m\ H_2O =$	ΔH -7.60				
	$M(H_2O)_x^{+3} + MS(H\ O)_{x-2}^{+2} + 2S(H_2O)_y^-$	ΔG 25.15	25	solubility of MS_3 in H_2O	-	64S
		ΔS -110				
	$M^{+3} + S^- = MS^{+2}$	3.49	25	HS aq. = $2\cdot10^{-5}$ (MS_3 sat. solution in H_2O) $\mu \to 0$	-	65S
Th^{+4}	$M^{+4} + S^- = MS^{+3}$	5.82	25	pH = 5.0, HS aq.$<2\cdot10^{-4}$	-	
	$MOH^{+3} + S^- = MPHS^{+2}$	5.06	25		-	
Ligand: ReO_4^- , A^-						
	$3\overline{(HS)_2} + H^+ + A^- = \overline{[(HS)_2]_3\cdot HA}$	-1.8	?	HA<3.0	HS 4.85 in CCl_4	61K
Ligand: NO_3^- , A^-						
H^+	$\overline{(HS)_2} + HA_{\text{unionized}} = \overline{(HS)_2\cdot HA}$	∼ -0.4	25	HA<12	HS 0.5 in toluene	61G
		∼ -0.7	25	9<HA<12	HS 0.05 in kerosene	
		-0.53 to-0.7	25	9 HA 12	HS 0.5 in toluene	
Be^{+2}	$M^{+2} + 2\overline{(HS)_2} = \overline{M(HS_2)_2} + 2H^+$	1.54	?	HA	HS in toluene	61H/59H
		1.55	?	HA 0.25	HS 0.5 in toluene	61H

METAL ION, M	REACTION	LOG K	TEMP.	CONDITIONS, AQ.	CONDITIONS, ORG.	REF.
Ligand: NO_3^- , A^- (cont.)						
Y^{+3}	$M^{+3} + 3\overline{(HS)_2} = \overline{M(HS_2)_3} + 3H^+$	3.3	25	0.1<HA<1.0	HS 0.1 in $CHCl_3$	57D
	$MS^{+2} + 2\overline{HS} = \overline{MS_3} + 2H^+$	2.12	22	HA<0.5	HS<0.1 in benzene	61S
Ce^{+3}		-0.45	20	HA<0.5	HS<0.1 in benzene	
Pr^{+3}	$M^{+3} + 3S^- + 3HS = \overline{M(HS_2)_3}$	15.0	20(?)	HA<1.18	HS<1.0 in CCl_4	62S,61Ka
Nd^{+3}		∿15.4	20(?)	HS<1.18	HS<1.0 in CCl_4	
Eu^{+3}		19.33	25	HA 0.1	HS<0.42 in $CHCl_3$	60Db
		16.80	25	HA 0.1	HS<0.5 in CCl_4	
		13.66	25	HA 0.1	HS<0.016 in n-hexane	
		14.94	25	HA 0.1	HS<0.5 in isopropyl ether	
		14.15	25	HA 0.1	HS<0.5 in MiBC	
		15.22	25	HA 0.1	HS<0.5 in MiBK	
		7.49	25	HA 0.1	HS<0.5 in TBP	
	$M^{+3} + 3S^- = \overline{MS_3}$	5.76	25	HA 0.1	HS<0.001 in MiBC	
	$M^{+3} + 6S^- + 3H^+ = \overline{M(HS_2)_3}$	16.6	25	HA 0.1	HS in hexane	62D
		18	25	HA 0.1	HS in isopropyl ether	

METAL ION, M	REACTION	LOG K	TEMP.	CONDITIONS, AQ.	CONDITIONS, ORG.	REF.
Ligand: NO_3^- , A^- (cont.)						
Eu^{+3}	$M^{+3} + 6S^- + 3H^+ = \overline{M(HS_2)_3}$	18.3	25	HA 0.1	HS in MiBK	62D
		19.8	25	HA 0.1	HS in CCl_4	
		22.4	25	HA 0.1	HS in $CHCl_3$	
	$M^{+3} + 3S^- + 2\overline{HS} = \overline{MS_3 \cdot (HS)_2}$	18.5	25	HA 0.1	HS 0.03-0.6 in $CHCl_3$	64K/60Db
	$M^{+3} + 2.5H_2S_2 = \overline{MS_3(HS)_2} + 3H^+$	1.02	25	HA 0.1	HS 0.03-0.6 in $CHCl_3$	64K/60Db
Yb^{+3}	$M^{+3} + 3S^- + 3\overline{HS} = \overline{M(HS_2)_3}$	18.6	20(?)	HA<1.18	HS<1.0 in CCl_4	62S,61Ka
Zr^{+4}	$MS_3^+ + HS = MS_4 + H^+$	2.60	20	HA 3.0, LiA 3.0	HS<0.075 in kerosene	62Sb
Th^{+4}	$M^{+4} + 4S^- + 2HS = \overline{MS_2(HS_2)_2}$	25.4 to 26.0	20	HA<0.95	HS<$2.5 \cdot 10^{-3}$ in CCl_4	63S
	$M^{+4} + A^- + 5HS = \overline{MAS_3(HS)_2} + 3H^+$	19.6	25	HA 1.0	HS<0.06 in $CHCl_4$	64D
	$M^{+4} + A^- + 4HS = \overline{MAS_3(HS)} + 3H^+$	13.4	25	$(Na^+,H^+)A^-$ 1.0 $[H^+] > 0.1$	HS<0.63 in MiBC	
	$M^{+4} + 2A^- + 2HS = \overline{MA_2S_2} + 2H^+$	6.6	25	$(Na^+,H^+)A^-$, $[H^+] > 0.1$	HS<0.04 in MiBC	
	$M^{+4} + A^- + 5HS = \overline{MAS_3(HS)_2} + 3H^+$	15.5	25	$H(A^-,Cl^-)$ 1.0	HS<0.02 in hexane	
		14.4	25	HA 1.0	HS<0.02 in hexane	
	$M^{+4} + 4A^- = \overline{MA_4}$	-2.0	25	HA 1.0	HS<$2 \cdot 10^{-3}$ in MiBC	

METAL ION, M	REACTION	LOG K	TEMP.	CONDITIONS, AQ.	CONDITIONS, ORG.	REF.
Ligand: NO_3^- , A^- (cont.)						
Nb^{+5}	$M(OH)_3\ S_3^- + 2HS + H^+ = \overline{MS_5} + 3H_2O$	2.45	20	HA 3.0, LiA 3.0	HS<0.1 in benzene	62D
UO_2^{++}	$M^{+2} + 3\overline{(HS)_2} = \overline{M(HS_2)_3} + 3H^+$	4.55			HS in $CHCl_3$	60H
	$M^{+2} + 2A^- + 2\overline{HS} = \overline{MA_2(HS)_2}$	6.76			HS in toluene	60Dc
	$M^{+2} + 2\overline{(HS)_2} = \overline{M(HS_2)_2} + 2H^+$	4.56		HA<4	HS<0.1 in toluene	61H
	$M^{+2} + \overline{2HS} = \overline{MS_2} + 2H^+$	4.69	20	LiA 2.0, HA 2.0	HS 0.1 in hydrogenated kerosene	62Sa
	$\overline{MS_2} + \overline{TBP} = \overline{MS_2 \cdot TBP}$	4.95	22	HA 2.0	HS<2.10^{-3},TBP<0.1 in benzene	62Sc
	$M^{+2} + A^- + S^- + \overline{TBP} = \overline{M \cdot A \cdot S \cdot TBP}$	4.80	?	HA 0.04	HS<0.02, TBP 0.95 in Amsco 125-90W	64H
	$2M^{+2} + 2A^- + 2S^- + 2\overline{TBP} = \overline{[MS_2][MA_2 \cdot 2TBP]}$	12.5	?	HA 0.04	HS 0.002, TBP<0.95 in Amsco 125-90W	
Pu^{+3}	$M^{+3} + 6S^- + 3H^+ = \overline{M(HS_2)_3}$	~17	25	HA 0.1	HS in dibutylether	62D
		~19	25	HA 0.1	HS in CCl_4	
Pu^{+4}	$M^{+4} + 2A^- + 2\overline{(H_2S_2)} = \overline{MA_2(HS_2)_2} + 2H^+$	9.23	?	$\mu = 4\ H^+(A^-, ClO_4^-)$	HS<$5.6 \cdot 10^{-3}$ in benzene	63D
	$M^{+4} + 2A^- + \overline{H_2S_2} + \overline{S'} = \overline{MA_2S_2 \cdot S'}$(?)	15.0	?	HA 1.0	HS<10^{-3}, S'<0.02	66Ka
	S' = Trihexylphosphine oxide					

METAL ION, M	REACTION	LOG K	TEMP.	CONDITIONS, AQ.	CONDITIONS, ORG.	REF.
Ligand: NO_3^- , A^- (cont.)						
Pu^{+4}	$\Sigma\, M(IV) + 2\overline{H_2S_2} = \overline{MA_2S_2(HS)_2} + 2H^+$	7.7	19-25	$\mu = 6\ (H^+, Na^+)A^-$	HS<0.05 in CCl_4	67Ka
	$\Sigma\, M(IV) + 2\overline{H_2S_2} = \overline{MAS_3(HS)} + 3H^+$	8.2		$\mu = 6\ (H^+, Na^+)A^-$	HS<0.05 in CCl_4	
	$MA^{+3} + 2\overline{H_2S_2} = \overline{MAS_3(HS)} + 3H^+$	8.4		$\mu = 1\ H^+(A^-, ClO_4^-)$	$HS<10^{-3}$ in CCl_4	
		7.8		HA 1.0	HS<0.002 in CCl_4	
	$MA^{+3} + 2.5\overline{H_2S_2} = \overline{MAS_3(HS)_2} + 3H^+$	10.6		$\mu = 1\ H^+(A^-, ClO_4^-)$	$HS<10^{-3}$ in CCl_4	
		10.0		HA 1.0	HS<0.002 in CCl_4	
Am^{+3}	$M^{+3} + 3S^- + 3HS = \overline{M(HS_2)_3}$	17.98	25	HA 0.1	HS<0.42 in $CHCl_3$	60Db
		15.33	25	HA 0.1	HS<0.50 in CCl_4	
		12.38	25	HA 0.1	HS<0.016 in n-hexane	
		14.25	25	HA 0.1	HS<0.5 in MiBK	
		14.02	25	HA 0.1	HS<0.5 in MiBC	
		13.72	25	HA 0.1	HS<0.5 in isopropylether	
		6.98	25	HA 0.1	HS<0.5 in TBP	
	$M^{+3} + 3S^- = \overline{MS_3}$	5.36	25	HA 0.1	HS<0.001 in MiBC	
	$M^{+3} + 3\overline{H_2S_2} = \overline{M(HS_2)_3} + 3H^+$	0.2	20	HA<1.0 The same value for A = ClO_4^-, CH_3COO^- and $HC_2O_4^-$	HS 30% in CCl_4	63Z

METAL ION, M	REACTION	LOG K	TEMP.	CONDITIONS, AQ.	CONDITIONS, ORG.	REF.
Ligand: PO_4^{-3} , A^{-3}						
H^+	$1.8\ H_3A + \overline{H_2S_2} = \overline{2HS \cdot 1.8\ H_3A}$	0.31	25	$\mu = 1.07\ (H^+, Na^+)\ ClO_4^-$ $H_3A < 3.76$	$HS < 0.002$ in kerosene	60T
	$2H_3A + \overline{H_2S_2} = \overline{2HS \cdot 2H_3A}$	0.2-0.3	25	$\mu = 1.07\ (H^+, Na^+)\ ClO_4^-$, $H_3A < 3.76$	$HS < 0.002$ in kerosene	
Ligand: SO_4^{-2} , A^{-2}						
H^+	$\overline{2R_3N \cdot HS} + 2H^+ + A^{-2} = \overline{(R_3N \cdot H)_2A} + \overline{2HS}$	2.16	?	H_2A 2.08	$HS < 0.1$, R_3N 0.1 in benzene	65D
	R_3N = Trioctylamine					
La^{+3}	$M^{+3} + S^- = MS^{+2}$	1.77	25	$H_2A < 0.12$ solubility of MS_3 in H_2O		64Sa
UO_2^{++}	$M^{+2} + 2S^- + \overline{2HS} = \overline{MS_2(HS)_2}$	15.6	25	0.1 H_2A	$HS < 0.06$, $TBP < 0.1$ in CCl_4	60D
	$M^{+2} + 2S^- + \overline{HS} + \overline{TBP} = \overline{MS_2HS \cdot TBP}$	12.6	25	0.1 H_2A	$HS < 0.06$, $TBP < 0.1$ in CCl_4	
	$\overline{MS_2(HS)_2} + \overline{TBP} = \overline{MS_2(HS) \cdot TBP} + \overline{HS}$	3.0	25	0.1 H_2A	$HS < 0.06$, $TBP < 0.1$ in CCl_4	
	$M^{+2} + 4HS = \overline{MS_2(HS)_2} + 2H^+$	8.35	25	0.1 H_2A	$HS < 0.26$ in hexane	66D
		8.56	25	0.1 H_2A	$HS < 0.26$ in hexane	67L/66D
		10.71	25	0.1 H_2A	$HS < 0.004$ in CCl_4	
	$M^{+2} + 2HS + \overline{TBP} = \overline{MS_2 \cdot TBP} + 2H^+$	4.6	25	0.1 H_2A	$HS < 10^{-3} + TBP > 2$ in hexane	

METAL ION, M	REACTION	LOG K	TEMP.	CONDITIONS, AQ.	CONDITIONS, ORG.	REF.
Ligand: SO_4^{-2}, A^{-2} (cont.)						
UO_2^{++}	$M^{+2} + 4HS + TBP = \overline{MS_2(HS)_2 \cdot TBP} + 2H^+$	11.51	25	0.1 H_2A	HS<0.004 + TBP<1.5 in hexane	66D
		11.38	25	0.1 H_2A		67L/66D
		12.79	25	0.1 H_2A	HS<$10^{-2.9}$ + TBP<0.055 in CCl_4	66D
	$M^{+2} + 3HS + \overline{TBP} = \overline{MS_2(HS).TBP} + 2H^+$	7.9	25	0.1 H_2A	HS<$10^{-2.9}$ + TBP<0.055 in hexane	
		7.9	25	0.1 H_2A	HS $10^{-2.8}$ + TBP in hexane	67L
	$M^{+2} + 2HS + \overline{TOPO} = \overline{MS_2 \cdot TOPO} + 2H^+$	8.18	25	0.1 H_2A	HS>10^{-4},TO'PO <0.055 in hexane	66D
		8.66	25	0.1 H_2A	$[HS]^2[TO'PO]^{-1} > 10^{-7.4}$ TO'PO<$6 \cdot 10^{-2}$ in CCl_4	
	$M^{+2} + 3HS + \overline{TOPO} = \overline{MS_2(HS) \cdot TOPO} + 2H^+$	11.95	25	0.1 H_2A	HS<$10^{-2.8}$ + TOPO<$4.7 \cdot 10^{-}$ in CCl_4	
Ligand: Cl^- , A^-						
Sc^{+3}	$\overline{MS_3} + 1.5\ \overline{H_2S_2} = \overline{M(HS_2)_3}$	∿34	20	HA<6	HS 0.001 in isobutylacetate	64Sb
Y^{+3}	$M^{+3} + 3\overline{(HS)_3} = \overline{M(SHS)_3} + 3H^+$	2.7	22	HA(?)<0.5	HS<1.5 in hexane(?)	65K/59D
La^{+3}		-2.52	22	HA(?)<5.0	HS<1.5 in hexane (?)	
Th^{+4}	$M^{+4} + 6HS = \overline{MS_4(HS)_2} + 4H^+$	22.5	25	$(H^+,Na^+)A^-$ 1.0 $[H^+]>0.1$	HS<0.06 in $CHCl_3$	64D
	$M^{+4} + S^- + 5HS = \overline{MS_4(HS)_2} + 3H^+$	23.5	25	(H^+,Na^+) A^- 1.0 $[H^+]>0.1$	HS<0.06 in $CHCl_3$	

METAL ION, M	REACTION	LOG K	TEMP.	CONDITIONS, AQ.	CONDITIONS, ORG.	REF.
Ligand: Cl^-, A^- (cont.)						
Po^{+4}	$MS_3^+ + \overline{(HS)_2} = \overline{MS_4 \cdot HS} + H^+$	1.40	?	HA 2.5	HS<0.8 in CCl_4	63A
	$M(HS)_2)_3^+ + \overline{(HS)_2} = \overline{M(HS_2)_4} + H^+$	1.34	?	HA 2.5	HS<0.8 in CCl_4	
Ligand: ClO_4^- , A^-						
Sc^{+3}	$M^{+3} + 3\overline{H_2S_2} = \overline{M(HS_2)_3} + 3H^+$	5.06	?	HA 0.1	HS<1 in dibutyl ether	62B/60Db
Y^{+3}	$M^{+3} + 6S^- + 3H^+ = \overline{M(HS_2)_3}$	19.7	25	HA 0.1	$1.4 \cdot 10^{-3}$<HS<0.13 in dibutyl ether	62D
	$M^{+3} + 3S^- + 2\overline{HS} = \overline{MS_3(HS)_2}$	20.64	20	$(H^+,Na^+)A^-$ $\mu = 1.0$	HS 10^{-4}-10^{-3} in $CHCl_3$	64K
	$M^{+3} + 2.5\overline{H_2S_2} = \overline{MS_3(HS)_2} + 3H^+$	3.12	20	$(H^+,Na^+)A^-$ $\mu = 1.0$	HS 10^{-4}-10^{-3} in $CHCl_3$	64K
La^{+3}	$M^{+3} + 3\overline{H_2S_2} = \overline{M(HS_2)_3} + 3H^+$	1.23	25	HA 0.1	$1.4 \cdot 10^{-3}$<HS<0.13 in dibutyl ether	62B/60Db
		-0.44	25	$\mu = 1.0$ $(H^+,Na^+)A^-$	HS 0.05 in toluene	66Kb
	$M^{+3} + 6S^- + 3H^+ = \overline{M(HS_2)_3}$	15.9	25	HA 0.1	0.01<HS<0.8 in dibutyl ether	62D
		18.0	25	HA 0.1	0.01<HS<0.8 in CCl_4	
Ce^{+3}	$M^{+3} + 3H_2S_2 = M(HS_2)_3 + 3H^+$	2.00	?	HA 0.1	HS<1.0 in dibutyl ether	62B/60Dc
		0.10	25	$\mu = 1.0(H^+,Na^+)A^-$	HS 0.05 in toluene	66Kb
	$M^{+3} + 6S^- + 3H^+ = \overline{M(HS_2)_3}$	16.5	25	HA 0.1	$6.4 \cdot 10^{-3}$<HS<0.64 in dibutyl ether	62D
		18.3	25	HA 0.1	$6.4 \cdot 10^{-3}$<HS<0.64 in CCl_4	

METAL ION, M	REACTION	LOG K	TEMP.	CONDITIONS, AQ.	CONDITIONS, ORG.	REF.
Ligand: ClO_4^- , A^- (cont.)						
Pr^{+3}	$M^{+3} + 3H_2S_2 = \overline{M(HS_2)_3} + 3H^+$	2.00	?	HA 0.1	HS<1.0 in dibutyl ether	62B/60D
	$M^{+3} + 6S^- + 3H^+ = \overline{M(HS_2)_3}$	16.6	25	HA 0.1	0.01<HS<0.8 in dibutyl ether	62D
Pm^{+3}	$M^{+3} + 3H_2S_2 = \overline{M(HS_2)_3} + 3H^+$	2.54	?	HA 0.1	HS<1.0 in dibutyl ether	62B/60D
		0.96	25	$\mu = 1.0\ (H^+, Na^+)A^-$	HS 0.05 in toluene 66Kb	
	$M^{+3} + 6S^- + 3H^+ = \overline{M(HS_2)_3}$	17.0	25	HA 0.1	0.01<HS<0.8 in dibutyl ether	62D
Sm^{+3}	$M^{+3} + 3H_2S_2 = \overline{M(HS_2)_3} + 3H^+$	2.84	?	HA 0.1	HS 1.0 in dibutyl ether	62B/60D
		1.51	25	$\mu = 1.0\ (H^+, Na^+)A^-$	HS 0.05 in toluene	66Kb
	$M^{+3} + 6S^- + 3H^+ = \overline{M(HS_2)_3}$	17.3	25	HA 0.1	0.003 HS<0.56 in dibutyl ether	62D
Eu^{+3}	$M^{+3} + \overline{3H_2S_2} = \overline{M(HS_2)_3} + 3H^+$	3.11	?	HA 0.1	HS 1.0 in dibutyl ether	62B/60D
		1.73	25	$\mu = 1.0\ (H^+, Na^+)A^-$	HS 0.05 in toluene	66Kb
	$M^{+3} + 6S^- + 3H^+ = \overline{M(HS_2)_3}$	17.7	25	HA 0.1	$6.4 \cdot 10^{-3}$<HS<0.64 in dibutyl ether	62D
		19.9	25	HA 0.1	$6.4 \cdot 10^{-3}$<HS<0.64 in CCl_4	
Gd^{+3}	$M^{+3} + \overline{3H_2S_2} = \overline{M(HS_2)_3} + 3H^+$	0.2	?	HA 0.1	HS 1.0 in dibutyl ether	62B/60Dc

METAL ION, M	REACTION	LOG K	TEMP.	CONDITIONS, AQ.	CONDITIONS, ORG.	RFF.
Ligand: ClO_4^- , A^- (cont.)						
Gd^{+3}	$M^{+3} + 6S^- + 3H^+ = M(HS_2)_3$	17.9	25	HA 0.1	0.003<HS<0.56 in dibutyl ether	62D
		20.1	25	HA 0.1	0.003<HS<0.56 in CCl_4	
Tb^{+3}	$M^{+3} + 3\overline{H_2S_2} = \overline{M(HS_2)_3} + 3H^+$	4.10	?	HA 0.1	HS<1.0 in dibutyl ether	62B/60Dc
		2.79	25	μ = 1.0 $(H,Na^+)A^-$	HS 0.05 in toluene	66Kb
	$M^{+3} + 6S^- + 3H^+ = \overline{M(HS_2)_3}$	18.7	25	HA 0.1	0.003<HS<0.56 in dibutyl ether	62D
	$M^{+3} + 2.5\overline{H_2S_2} = \overline{MS_3(HS)_2} + 3H^+$	1.83	20	μ = 1.0 $(H^+,Na^+)A^-$	HS 10^{-4}-10^{-3} in $CHCl_3$	64K
	$M^{+3} + 3S^- + 2\overline{HS} = \overline{MS_3(HS)_2}$	19.29	20	μ = 1.0 $(H^+,Na^+)A^-$	HS 10^{-4}-10^{-3} in $CHCl_3$	64K
Dy^{+3}	$M^{+3} + 3\overline{H_2S_2} = \overline{M(HS_2)_3} + 3H^+$	3.19	25	μ = 1.0 $(H^+,Na^+)A^-$	HS 0.05 in toluene	66Kb
Ho^{+3}		5.00	?	HA 0.1	HS<1.0 in dibutyl ether	62B/60Dc
		3.57	25	μ = 1.0 $(H^+,Na^+)A^-$	HS 0.05 in toluene	66Kb
	$M^{+3} + 6S^- + 3H^+ = \overline{M(HS_2)_3}$	19.5	25	HA 0.1	$4\cdot10^{-4}$<HS<0.06 in dibutyl ether	62D
		21.7	25	HA 0.1	$4\cdot10^{-4}$<HS<0.06 in CCl_4	
Er^{+3}	$M^{+3} + 3\overline{H_2S_2} = \overline{M(HS_2)_3} + 3H^+$	3.95	25	μ = 1.0 $(H^+,Na^+)A^-$	HS 0.05 in toluene	66Kb

METAL ION, M	REACTION	LOG K	TEMP.	CONDITIONS, AQ.	CONDITIONS, ORG.	REF.
Ligand: ClO_4^- , A^- (cont.)						
Tm^{+3}	$M^{+3} + 6S^- + 3H^+ = \overline{M(HS_2)_3}$	20.5	25	HA 0.1	4.10^{-4}<HS<0.06 in dibutyl ether	62D
	$M^{+3} + 3S^- + 2\overline{HS} = \overline{MS_3(HS)_2}$	21.14	20		HS 10^{-4}-10^{-3} in $CHCl_3$	64K
	$M^{+3} + 2.5\overline{H_2S_2} = \overline{MS_3(HS)_2} + 3H^+$	3.63	20	$\mu = 1.0$ $(H^+,Na^+)A^-$	HS 10^{-4}-10^{-3} in $CHCl_3$	64K
	$M^{+3} + 3\overline{H_2S_2} = \overline{M(HS_2)_3} + 3H^+$	4.59	25	$\mu = 1.0$ $(H^+,Na^+)A^-$	HS 0.05 in toluene	66Kb
		6.20	?	HA 0.1	HS<1.0 in dibutyl ether	62B/60Dc
Yb^{+3}		5.00	25		HS 0.05 in toluene	66Kb
Lu^{+3}	$M^{+3} + 6S^- + 3H^+ = \overline{M(HS_2)_3}$	21.4	25	HA 0.1	4.10^{-4}<HS<0.06 in dibutyl ether	62D
		23.7	25	HA 0.1	4.10^{-4}<HS<0.06 in CCl_4	
	$M^{+3} + 3\overline{H_2S_2} = \overline{M(HS_2)_3} + 3H^+$	5.50	25	$\mu = 1.0$ $(H^+,Na^+)A^-$	HS 0.05 in toluene	66Kb
		7.70	?	HA 0.1	HS<1.0 in dibutyl ether	62B/60Dc
Hf^{+4}	$M^{+4} + A^- + 5HS = \overline{MAS(HS_2)_2} + 3H^+$	20.3	19	$\mu = 2.00$ $(H^+,Na^+)A^-$	HS<5.10^{-4} in CCl_4	67N
		20.4			HS<5.10^{-4} in benzene	
		23.2			HS<5.10^{-4} in $CHCl_3$	
Th^{+4}	$M^{+4} + a^- + 5HS = \overline{MAS_3(HS)_2} + 3H^+$	19.6	25	$(Na^+,H^+)A^-$ 1.0, $[H^+]$ 0.1	HS<0.1 in $CHCl_3$	64D

METAL ION, M	REACTION	LOG K	TEMP.	CONDITIONS, AQ.	CONDITIONS, ORG.	REF.
Ligand: ClO_4^- , A^- (cont.)						
Th^{+4}	$M^{+4} + A^- + 5HS = \overline{MAS_3(HS)_2} + 3H^+$	19.5	25	$(Na^+,H^+)A^-$ 1.0	HS<0.25, TBP<2 $CHCl_3$	
UO_2^{++}	$M^{+2} + 2S^- + 2\overline{HS} = \overline{MS_2(HS)_2}$	15.22	25	$\mu = 1$ $(Na^+,H^+)A^-$	HS<0.5 in $CHCl_3$	65K/59D
	$M^{+2} + 2\overline{H_2S_2} = \overline{M(HS)_2S_2} + 2H^+$	3.58	25	$\mu = 1$ $(Na^+,H^+)A^-$	HS<0.5 in $CHCl_3$	59D
		4.50	25	$\mu = 1$ $(Na^+,H^+)A^-$	HS<0.5 in MiBK	
	$\overline{MS_2} + 2\overline{HS} = \overline{MS_2(HS)_2}$	7.76	25	———	HS<0.5 in MiBK	
		4.78	25	———	HS<0.005 in MiBK	69D
	$2M^{+2} + 2\overline{H_2S_2} = \overline{M_2S_4} + 4H^+$	0.35	25	HA 0.1	HS 0.04 in $CHCl_3$	59D
	$M^{+2} + 2\overline{H_2S_2} = \overline{M(HS_2)_2} + 2H^+$	4.56	?	?	HS in toluene	62B/60H
Am^{+3}	$M^{+3} + 6S^- + 3H^+ = \overline{M(HS_2)_3}$	16.5	25	HA<0.5	$5\cdot10^{-3}$ HS<0.33 in dibutyl-ether	62D
Cm^{+3}		16.5	25	HA<0.5	$5\cdot10^{-3}$<HS<0.33 in dibutyl-ether	
Cf^{+3}		18.1	25	HA 0.1	$5\cdot10^{-3}$<HS<0.33 in dibutyl-ether	
Es^{+3}		17.9	25	HA 0.1	HS 0.02 in dibutylether	
Po^{+4}	$MS_3^+ + \overline{(HS)_2} = \overline{MS_4\cdot HS} + H^+$	0.93	?	HA 3.0	HS<0.8 in CCl_4	63A

REFERENCES

43K Kumler, W. D. and Eiler, J. J., J. Am. Chem. Soc., 65, 2355 (1943)

57D Dyrssen, D., Acta Chem. Scand., 11, 1277 (1957)

57Da Dyrssen, D., Acta Chem. Scand., 11, 1771 (1957)

57P Peppard, D. F., Mason, G. W. and Moline, S. W., J. Inorg. Nucl. Chem., 5, 141 (1957)

58D Dewandre, A., Mémoire de Licence, Université de Liège (1958)

59D Dyrssen, D. and Krasovec, F., Acta Chem. Scand., 13, 561 (1959)

59Da Dreze, Ph., Bull. Soc. Chim. Belges, 68, 674 (1959)

59H Hardy, C. J. and Greenfield, B. F., (unpublished work (1959))

59Ha Hardy, C. J. and Scargill, D., J. Inorg. Nucl. Chem., 11, 128 (1959)

60H Hardy, C. J.,British Report AERE 3267 (1960)

60D Dyrssen, D. and Kuca, L., Acta Chem. Scand., 14, 1945 (1960)

60Da Dyrssen, D. and Liem, D. H., Acta Chem. Scand., 14, 1091 (1960)

60Db Dyrssen, D. and Liem, D. H., Acta Chem. Scand., 14, 1100 (1960)

60Dc Duyckaerts, G., Dreze, Ph. and Simon, A., J. Inorg. Nucl. Chem., 13, 332 (1960)

60T Thamer, B. J., J. Phys. Chem., 64, 694 (1960)

61G Greenfield, B. F. and Hardy, C. J., J. Inorg. Nucl. Chem., 21, 359 (1961)

61H Hardy, C. J., J. Inorg. Nucl. Chem., 21, 348 (1961)

61Ha Hardy, C. J., Greenfield, B. F. and Scargill, D., J. Chem. Soc., 174 (1961)

61K Kertes, A. S. and Beck, A., J. Chem. Soc., 5046 (1961)

61Ka Kriss, E. E. and Sheka, Z. A., Dokl. Akad. Nauk SSSR, 138, 846 (1961)

61S Shevchenko, V. B. and Smelov, V. S., Zh. Neorg. Khim., 6, 732 (1961)

62B Baes, C. F. Jr., J. Inorg. Nucl. Chem., 24, 707 (1962)

62D Duyckaerts, G. and Dreze, Ph., Bull. Soc. Chim. Belges, 71, 306 (1962)

62S Sheka, Z. A. and Kriss, E. E., Zh. Neorg. Khim., 7, 658 (1962)

62Sa Shevchenko, V. B. and Smelov, V. S., Ekstrakstziya Teoriya, Primenenie, Apparatura, Sb Statie, 2, 219 (1962)

62Sb Shevchemko, V. B. and Smelov, V. S., Ekstrakstziya Teoriya, Primenenie, Apparatura, Sb Statie, 2, 257 (1962)

62Sc Shevchenko, V. B. and Smelov, A. V., Strachova, Ekstrakstziya Teoriya, Primenie, Apparatura, Sb Statie, 2, 179 (1962)

63A Ampelogova, N. I., Radiokhimiya, 5, 562 (1963)

63D Dreze, Ph. and Duyckaerts, G., EURATOM Report 436 f (1963)

63K Kriss, E. E., Zh. Neorg. Khim., 8, 1505 (1963)

63S Sheka, Z. A. and Sinyavskaya, E. I., Radiokhimiya, 5, 485 (1963)

63Z Zemlyanukhin, V. I., Savoskina, G. P. and Pushlenkov, M. F., Radiokhimiya, 5, 674 (1963)

64D Dyrssen, D. and Liem, D. H., Acta Chem. Scand., 18, 224 (1964)

64H Hahn, H. T. and Van der Wall, E. M., J. Inorg. Nucl. Chem., 26, 191 (1964)

64K Krasovec, F. and Klofutar, C., Yugoslave Report NIJS - R-440 (1964)

64S Sheka, Z. A. and Sinyavskaya, E. I., Zh. Neorg. Khim., 9, 1974 (1964)

64Sa Sheka, Z. A. and Sinyavskaya, E. I., Zh. Neorg. Khim., 9, 2244 (1964)

64Sb Samodelov, A. P., Radiokhimiya, 6, 286 (1964)

65D Deptula, C. and Minc, S., Nukleonika, 10, 421 (1965)

65K Krasovec, F. and Klofutar, C., J. Inorg. Nucl. Chem., 27, 2437 (1965)

65S Sheka, Z. A. and Sinyavskaya, E. I., Radiokhimiya, 7, 596 (1965)

65Sa Sheka, Z. A. and Sinyavskaya, E. I., Zh. Neorg. Khim., 10, 394 (1965)

65Sb Sheka, Z. A. and Sinyavskaya, E. I., Zh. Neorg. Khim., 10, 259 (1965)

65U Ul'yanov, V. S. and Sviridova, R. A., Radiokhimiya, 7, 538 (1965)

66D Liem, D. H. and Dyrssen, D., Acta Chem. Scand., 20, 272 (1966)

66K Kiss, A. and Hegedus, A. J., Mikrochim. Acta, 771 (1966)

66Ka Kuca, L., Coll. Czech. Chem. Commun., 31, 4064 (1966)

66Kb Kolarik, Z. and Pankova, H., J. Inorg. Nucl. Chem., 28, 2325 (1966)

66S Sinyavskaya, E. I. and Sheka, Z. A., Zh. Neorg. Khim., 11, 1210 (1966)

67K Kuca, L., Coll. Czech. Chem. Commun., 32, 729 (1967)

67Ka Kuca, L., Coll. Czech. Chem. Commun., 32, 298 (1967)

67Kb Kuca, L., Coll. Czech. Chem. Commun., 32, 288 (1967)

67Kc Kolarik, Z., Coll. Czech. Chem. Commun., 32, 311 (1962)

67Kd Kuca, L., Coll. Czech. Chem., Commun., 32, 720 (1967)

67L Liem, D. H., in "Solvent Extraction Chemistry", Eds. Dyrssen, D., Liljenzin, J. O., Rydberg, J., North-Holland, Amsterdam, p. 264 (1967)

67N Navratil, O., in "Solvent Extraction Chemistry", Eds. Dyrssen, D., Liljenzin, J. O., Rydberg, J., North-Holland, Amsterdam, p. 256 (1967)

68L Liem, D. H., Acta Chem. Scand., 22, 753 (1968)

METAL ION, M	REACTION	LOG K	TEMP.	CONDITIONS, AQ.	CONDITIONS, ORG.	REF.
Extractant: Diethylhexylphosphoric acid (H)DE-HP						
Distribution Constants						
	$HS = \overline{HS}$	3.42	20	$(H^+,Na^+)ClO_4^-$ 0.1	HS<0.5 in n-octane	63U
		3.3		$(H^+,Na^+)ClO_4^-$ 0.1	HS<0.5 in n-octane	67K/63U
		1.30	25	$HClO_4$ 0.2	HS 10^{-4}-1.0 in benzene	66Sa
		1.22	25	$HClO_4$ 0.2	HS 10^{-4}-1.0 in CCl_4	66Sa
Dissociation						
	$\overline{HS} = \overline{H^+} + \overline{S^-}$	-3.35	?	——	HS 0.05 in ethanol 98%	65D
		-3.2	25	——	HS in ethanol 75%	55B
	$HS = H^+ + S^-$	-1.40	20	$(H^+,Na^+)ClO_4^-$ 0.1	HS<0.5 in n-octane	63U
		-2.85	25	$HClO_4$ 0.2		66Sa
		-1.72	20	$(H^+,Na^+)ClO_4^-$ 0.1	HS<0.5 in n-octane	69S
		ΔH-9.15				
Dimerization						
	$2\overline{HS} = \overline{(HS)_2}$	<4.48	22	——	HS<7.10^{-4} in octane	62Ba

METAL ION, M	REACTION	LOG K	TEMP.	CONDITIONS, AQ.	CONDITIONS, ORG.	REF.
Dimerization (cont.)						
	$2\overline{HS} = \overline{(HS)_2}$	< 2.9	25	$HClO_4$ 0.2	HS 10^{-4}-1.0 in benzene	66Sa
		2.0	25	$HClO_4$ 0.2	HS 10^{-4}-1.0 in CCl_4	66Sa
		4.47	20	$(H^+,Na^+)ClO_4^-$ 0.1	HS<0.5 in n-octane	63U
		4.70	20	$(H^+,Na^+)ClO_4^-$ 0.1	HS<0.5 in n-octane	67K/63U
Reaction with neutral organophosphorus compounds						
	$\overline{TBP} + \overline{HS} = \overline{TBP(HS)}$	0.78	25	$HClO_4$ 0.2	HS 0.0014, TBP<1.3 in benzene	66Sa
		1.12	25	$HClO_4$ 0.2	HS 0.0014, TBP<1.3 in CCl_4	66Sa
	$\overline{TBP} + \frac{1}{2}\overline{(HS)_2} = \overline{TBP \cdot HS}$	0.0	25	H_2SO_4 1.5	HS<0.4, TBP<0.5 in kerosene	63B/59B
		-0.22	25	H_2SO_4 1.5	HS<0.4, TBP<0.5 in hexane	
		-0.52	25	H_2SO_4 1.5	HS<0.4, TBP<0.5 in CCl_4	
		-0.6	25	H_2SO_4 1.5	HS<0.4, TBP<0.5 in benzene	
	$\overline{TBP} + \overline{(HS)_2} = \overline{TBP \cdot (HS)_2}$	0.60	25	H_2SO_4 1.5	HS<0.4, TBP<0.5 in kerosene	
		0.4	25	H_2SO_4 1.5	HS<0.4, TBP<0.5 in hexane	
		0.08	25	H_2SO_4 1.5	HS<0.4, TBP<0.5 in CCl_4	
		0.0	25	H_2SO_4 1.5	HS<0.4, TBP<0.5 in benzene	
		>1.17	?	HNO_3 0.05	HS 0.1, TBP<0.1 in CCl_4	66Z

METAL ION, M	REACTION	LOG K	TEMP.	CONDITIONS, AQ.	CONDITIONS, ORG.	REF.
Reaction with neutral organophosphorus compounds (cont.)						
	$\overline{DBB'P} + \frac{1}{2}\overline{(HS)_2} = \overline{DBB'P \cdot HS}$	0.17	25	H_2SO_4 1.5	HS<0.4, DBP'P<0.5 in kerosene	63B/59B
	$\overline{DBB'P} + \overline{(HS)_2} = \overline{DBB'P \cdot (HS)_2}$	0.78	35	H_2SO_4 1.5	HS<0.4, DBB'P<0.5 in kerosene	63B/59B
		∿1.30	?	——	HS<0.01, DBB'P<1.0 in cyclohexane	63Z
	$\overline{BDB'P} + \frac{1}{2}\overline{(HS)_2} = \overline{BDB'P \cdot HS}$	0.65	25	H_2SO_4 1.5	HS<0.4, BDB'P<0.4 in kerosene	63B/59B
	$\overline{BDB'P} + \overline{(HS)_2} = \overline{BDB'P \cdot (HS)_2}$	1.25	25	H_2SO_4 1.5	HS<0.4, BDB'P<0.5 in kerosene	
	$\overline{TB'PO} + \frac{1}{2}\overline{(HS)_2} = \overline{TB'PO \cdot HS}$	0.92		H_2SO_4 1.5	HS<0.4, TB'PO<0.5 in kerosene	
	$\overline{TB'PO} + \overline{(HS)_2} = \overline{TB'PO \cdot (HS)_2}$	1.52				
	$\overline{TO'PO} + \frac{1}{2}\overline{(HS)_2} = \overline{TO \cdot HS}$	1.38	22	——	$(HS)_2$ + TO'PO 0.1 and 0.2 in octane	62Ba
		1.48	?	——	TO'PO + HS in kerosene	63I/58Ba
	$\overline{TO'PO} + \overline{(HS)_2} = \overline{TO'PO \cdot (HS)_2}$	2.0	22	——	$(HS)_2$ + TO'PO 0.1 and 0.2 in octane	62Ba
		∿2	?	——	TO'PO + HS in kerosene	63I/58Ba
		∿2.0	?	——	HS<0.01, TO'PO<0.45 in cyclohexane	63Z
Ligand: NO_3^- , A^-						
H^+	$H^+ + A^- + \overline{(HS)_2} = \overline{HA(HS)_2}$	-1.16	?	HA → 0	$(HS)_2$ → 0 in CCl_4	68S
Na^+	$M^+ + \overline{(HS)_2} = \overline{MHS_2} + H^+$	-2.90	?	HA ?	HS in Amsco 80-125	62B/60L
		-3.9	?	HA ?	HS in toluene	

METAL ION, M	REACTION	LOG K	TEMP.	CONDITIONS, AQ.	CONDITIONS, ORG.	REF.
Ligand: NO_3^- , A^- (cont.)						
K^+	$M^+ + \overline{(HS)_2} = \overline{MHS_2} + H^+$	-2.94	?	HA ?	HS in Amsco 80-125	62B/60L
		-3.9	?	HA ?	HS in toluene	
Rb^+		-2.95	?	HA ?	HS in Amsco 80-125	
		-4.0	?	HA ?	HS in toluene	
Cs^+		-3.25	?	HA ?	HS in Amsco 80-125	
		-4.4	?	HA ?	HS in toluene	
Be^{++}	$M^{+2} + 2\overline{(HS)_2} = M\overline{(HS_2)_2} + 2H^+$	1.28		HA 0.25	HS 0.5 in toluene	61H
		1.28	?	HA ?	HS in toluene	61Ha/59H
		1.28	?	HA ?	HS in toluene	62B/60H
Ca^{++}		-1.2	?	HA ?	HS in Amsco 80-125	62B/60L
		-1.8	?	HA ?	HS in toluene	
Sr^{++}		-3.2	?	HA ?	HS in Amsco 80-125	
		-3.8	?	HA ?	HS in toluene	
Ba^{++}		-3.9	?	HA ?	HS in Amsco 80-125	
		-4.5	?	HA ?	HS in toluene	

METAL ION, M	REACTION	LOG K	TEMP.	CONDITIONS, AQ.	CONDITIONS, ORG.	REF.
Ligand: NO_3^- , A^- (cont.)						
Y^{+3}	$M^{+3} + 3\overline{(HS)_2} = \overline{M(HS_2)_3} + 3H^+$	1.63	25	μ = 1.00 $(H^+,Na^+)A^-$	HS 0.05 in toluene	66K
Pm^{+3}		-1.73	25			
Sm^{+3}		-0.95	25			
Eu^{+3}		-0.83	25			
Tb^{+3}		0.40	25			
Dy^{+3}		0.88	25			
Ho^{+3}		1.25	25			
Er^{+3}		1.75	25			
Tm^{+3}		2.27	25			
Yb^{+3}		2.77	25			
Lu^{+3}		2.94	25			
Th^{+4}	$M^{+4} + 3\overline{(HS)_2} = \overline{MS_2(HS_2)_2} + 4H^+$	6.90	?	μ = 1.0 $H^+(A^-,ClO_4^-)$	HS 0.04 in toluene	60P
	$MA^{+3} + 3\overline{(HS)_2} = \overline{MA(HS_2)_3} + 3H^+$	4.65	?	μ = 1 $H^+(A^-,ClO_4^-)$	HS 0.04 in toluene	
	$mMA^{+3} + (2m+1)\overline{(HS)_2}$	ΔH 2.20	10-50	HA 1.0	HS 0.03 in kerosene	67S
	$= \overline{(MA)_m \cdot S_{2(2m+1)}H_{m+2}} + 3mH^+$ (m > 1)	ΔH 3.10	10-50	HA 1.0	HS 0.05 in kerosene	

METAL ION, M	REACTION	LOG K	TEMP.	CONDITIONS, AQ.	CONDITIONS, ORG.	REF.
Ligand: NO_3^- , A^- (cont.)						
Th^{+4}	$M^{+4} + 4A^- + \overline{(HS)_2} = \overline{MA_4(HS)_2}$	ΔH 4.50	10-50	HA 6.0	HS 0.03 in kerosene	67S
		ΔH 6.4	10-50	HA 6.0	HS 0.05 in kerosene	
UO_2^{++}	$M^{+2} + 2\overline{HS} = \overline{MS_2} + 2H^+$	1.93		μ = 0.25 HA	HS<0.5 in n-amylalcohol	64M
	$M^{+2} + 4\overline{HS} = \overline{M(HS_2)_2} + 2H^+$	3.4				
	$M^{+2} \quad 2\overline{(HS)_2} = \overline{M(HS_2)_2} + 2H^+$	3.84	?	HA → 0	HS → 0 in CCl_4	68S
		4.56	?	HA < 4.0	HS<0.1 in toluene	61Ha
		2.95	22	μ = 1.0 $(H^+,Na^+)A^-$	HS<0.1 in toluene	62P
		ΔG 5.31	25	μ = 0.75 HA 0.5	HS 0.15 in CCl_4	63M
		ΔH 0.80	25			
		ΔS 19.1	25			
		ΔH 2.46	10-50	HA 0.5	HS 0.1 in kerosene	63S
	$M^{+2} + 2A^- + 2\overline{(HS)_2} = \overline{MA_2 \cdot 2HS}$	ΔH 3.69	10-50	HA 0.5	HS 0.1 in kerosene	
		2.8	?	HA 3.0-9.0	$(HS)_2$ 0.075-0.25 in CCl_4	68S
Am^{+3}	$M^{+3} + 3\overline{HS} = \overline{MS_3} + 3H^+$	-0.57	22	HA<1.0	HS<0.5 in isooctane	64G
	$M^{+3} + 3\overline{(HS)_2} + \overline{TBP} = \overline{M(HS_2)_3 \cdot TBP} + 3H^+$	4.30	?	HA 0.05	HS 0.1, TBP<1.0 in CCl_4	66Z
AmO_2^{++}	$M^{+2} + 2\overline{(HS)_2} + \overline{TBP} = \overline{MS_3^- \cdot TBP \cdot HSH^+} + 2H^+$	6.17	?	HA 0.05	HS<0.03, TBP<1.0 in CCl_4	

METAL ION, M	REACTION	LOG K	TEMP.	CONDITIONS, AQ.	CONDITIONS, ORG.	REF.
Ligand: NO_3^- , A^- (cont.)						
Co^{++}	$M^{+2} + 2\overline{(HS)_2} = \overline{M(HS_2)_2} + 2H^+$	-3.4	?	HA ?	HS in Amsco 80-125	62B/60L
		-4.0		HA	HS in toluene	
Ag^+	$M^+ + \overline{(HS)_2} = \overline{M(HS_2)} + H^+$	-2.7		HA	HS in toluene	
		-1.69		HA	HS in Amsco 80-125	
Zn^{++}	$M^{+2} + 2\overline{(HS)_2} = \overline{M(HS_2)_2} + 2H^+$	-1.8		HA	HS in toluene	
		-1.2		HA	HS in Amsco 80-125	
Cd^{++}		-1.8		HA	HS in Amsco 80-125	
		2.4		HA	HS in toluene	
Ligand: SO_4^{-2}, A^{-2}						
H^+	$2\overline{R_3NHS} + 2H^+ + A^{-2} = \overline{(R_3NH)_2A} + 2\overline{HS}$	1.83	?	$H_2A<2.08$	HS 0.1, R_3N 0.1 in benzene	65D
	R_3N = Trioctyl amine					
Th^{+4}	$M^{+4} + 4\overline{(HS)_2} = \overline{M(HS_2)_4} + 4H^+$	ΔH 8.04	10-50	H_2A 0.25	HS 0.1 in kerosene	65Sa
		ΔH 8.11	10-50	H_2A 1.0	HS 0.1 in kerosene	
UO_2^{++}	$M^{+2} + 2\overline{(HS)_2} = \overline{M(HS_2)_2} + 2H^+$	ΔH 1.71	10-50	H_2A 0.15	HS 0.1 in kerosene	62S
		ΔH 2.42	10-50	H_2A 1.0	HS 0.1 in kerosene	

F

METAL ION, M	REACTION	LOG K	TEMP.	CONDITIONS, AQ.	CONDITIONS, ORG.	REF.
Ligand: SO_4^{-2}, A^{-2} (cont.)						
UO_2^{++}	$M^{+2} + 2\overline{(HS)_2} = \overline{M(HS_2)_2} + 2H^+$	ΔG -2.77	25	H_2A 0.25	HS 0.15 in $CHCl_3$	63M
		ΔH 1.40				
		ΔS 14.0				
		ΔG -2.83			HS 0.15 in n-octanol	
		ΔH -4.05				
		ΔS -4.11				
		ΔG -3.75			HS 0.15 in MiBK	
		ΔH -0.187				
		ΔS 12.0				
		ΔG -3.86			HS 0.15 in benzene	
		ΔH 0.363				
		ΔS 14.2				
		ΔG -3.92			HS 0.15 in CCl_4	
		ΔH 0.76				
		ΔS 15.7				

METAL ION, M	REACTION	LOG K	TEMP.	CONDITIONS, AQ.	CONDITIONS, ORG.	REF.
Ligand: SO_4^{-2} , A^{-2} (cont.)						
UO_2^{++}	$M^{+2} + 2\overline{(HS)_2} = \overline{M(HS_2)} + 2H^+$	ΔG -3.97	25	H A 0.25	HS 0.15 in chlorobenzene	63M
		ΔH 0.003				
		ΔS 13.3				
		ΔG -4.15			HS 0.15 in isopropylether	
		ΔH -0.23				
		ΔS 13.1				
		ΔG -4.83			HS 0.15 in kerosene	
		ΔH -0.57				
		ΔS 14.3				
		ΔG -4.29			HS 0.15 in isoamylacetate	
		ΔH -4.61				
		ΔS -1.1				
		ΔG -4.69			HS 0.15 in cyclohexane	
		ΔH -1.30				
		ΔS 11.4				

METAL ION, M	REACTION	LOG K	TEMP.	CONDITIONS, AQ.	CONDITIONS, ORG.	REF.
Ligand: SO_4^{-2}, A^{-2} (cont.)						
UO_2^{++}	$M^{+2} + 2(HS)_2 = M(HS_2)_2 + 2H^+$	ΔG -4.90	25	H_2A 0.25	HS 0.15 in n-hexane	63M
		ΔH -1.33	25	H_2A		
		ΔS 12.0				
		ΔG -4.75		μ = 0.75, H_2A 0.25	HS 0.15 in CCl_4	
		ΔH 0.72				
		ΔS 18.04				
	$M^{+2} + \overline{(HS)_2} + 2\overline{TBP} = \overline{MS_2 \cdot 2TBP} + 2H^+$	ΔH 51.0	10-50	H_2A 0.15	HS 0.1, TBP 0.1 in kerosene	64S
		ΔH 50.0	10-50	H_2A 1.0	HS 0.1, TBP 0.1 in kerosene	
	$\overline{M(HS_2)_2} + \overline{TBP} = \overline{M(HS_2)_2 \cdot TBP}$	2.17	25	H_2SO_4 1.5	HS<0.5, TBP<0.5 in kerosene	63B/59B
		1.84			HS<0.5, TBP<0.5 in hexane	
		1.60			HS<0.5, TBP<0.5 in CCl_4	
		1.20			HS<0.5, TBP<0.5 in benzene	
	$\overline{M(HS_2)_2} + \overline{DBB'P} = \overline{M(HS_2)_2 \cdot DBB'P}$	2.78	33		HS 0.5, DBB'P 0.5 in kerosene	
	$\overline{M(HS_2)_2} + \overline{BDB'P} = \overline{M(HS_2)_2 . BDB'P}$	3.78	25		HS<0.5, BDB'P<0.5 in kerosene	
	$\overline{M(HS_2)_2} + \overline{TB'PO} = \overline{M(HS_2)_2 \cdot TB'PO}$	4.28	25		HS<0.5, TB'PO<0.5 in kerosene	

METAL ION, M	REACTION	LOG K	TEMP.	CONDITIONS, AQ.	CONDITIONS, ORG.	REF.
Ligand: SO_4^{-2}, A^{-2} (cont.)						
UO_2^{++}	$\overline{MS_2(HS)_2} + \overline{TO'PO} = \overline{MS_2(HS_2)(TO'PO)}$	<3.85	?	H_2A 0.5	HS 0.1, TO'PO<0.02 in kerosene	63I
	$\overline{MS_2(HS)_2} + \overline{TO'PO} = \overline{MS_2(HS)\cdot TO'PO} + 0.5\overline{(HS)_2}$	<3.13	?	H_2A	HS 0.1, TO'PO<0.02 in kerosene	
	$\overline{MS_2(HS)\ (H_2O)_m} + \overline{TO'PO} = \overline{MS_2(HS)_2(H_2O)\cdot TO'PO} + (m-1)H_2O$	<1.74	?	H A ?	HS 0.1, TO'PO 0.02 in kerosene	
Al^{+3}	$M^{+3} + 3\overline{HS} = \overline{MS_3} + 3H^+$	-1.16	20	$M_2A_3\cdot 18\ H_2O$ pH <1.7	HS 0.485 in hydrogenated kerosene	65B
Ligand: Cl^- , A^-						
Na^+	$M^+ + \overline{(HS)_2} = \overline{MHS_2} + H^+$	-3.84	21	HA<2.0	HS 50% in toluene	62B/60K
K^+		-3.88				
Rb^+		-3.98				
Cs^+		-4.43				
Mg^{++}	$M^+ + 2\overline{(HS)_2} = \overline{M(HS_2)} + 2H^+$	-4.45 (?)				
Ca^{++}	$M^{+2} + 3\overline{(HS)_2} = \overline{M[(HS)_2]_2(HS)_2} + 2H^+$	-3.16	?	μ = 1.0 $(H^+,Na^+)A^-$	HS 0.3 in xylene	62Pa
Sr^{++}		-5.75 to-6.25	?	μ = 1.0 $(H^+,Na^+)A^-$	HS 0.3 in xylene	
	$M^{+2} + 2\overline{(HS)_2} = \overline{M(HS_2)_2} + 2H^+$	-4.47	21	HA<2.0	HS 50% in toluene	62B/60K
Ba^{++}		-3.6	?	HA<0.1	HS 1.5 in toluene	62B/57P
		-3.9	?	HA<0.1	HS 1.5 in toluene	
		-4.6	21	HA<2.0	HS 50% in toluene	62B/60K

METAL ION, M	REACTION	LOG K	TEMP.	CONDITIONS, AQ.	CONDITIONS, ORG.	REF.
Ligand: Cl^- , A^- (cont.)						
Sc^{+3}	$M^{+3} + 3\overline{HS} = \overline{MS_3} + 3H^+$	3.94	22	μ = 0.5 HA	HS<0.3 in decanol	64M
	$M^{+3} + 6\overline{HS} = \overline{M(HS_2)_3} + 3H^+$	7.48	22	μ = 0.5 HA	HS<0.3 in decanol	
Y^{+3}	$M^{+3} + 3\overline{(HS)_2} = \overline{M(HS_2)_3} + 3H^+$	2.70	22	HA 5(?)	HS<1.5 in hexane (?)	61Ha/57P
	$M^{+3} + 3\overline{HS} = \overline{MS_3} + 3H^+$	2.16	22	μ = 0.025 HA	HS<0.3 in decanol	
	$M^{+3} + 6\overline{HS} = \overline{M(HS_2)_3} + 3H^+$	4.54	22	μ = 0.025 HA	HS<0.3 in decanol	
La^{+3}	$M^{+3} + 3\overline{(HS)_2} = \overline{M(HS_2)_3} + 3H^+$	-2.52	22	HA 5 (?)	HS 1.5 in hexane (?)	
		-2.52	22	HA 0.5	HS 1.5 in toluene	61H/58Pa
Ce^{+3}		-2.22	22	HA<0.5	HS<0.9 in toluene	62B/57Pa
Pr^{+3}		-1.92	22	HA<0.5	HS<0.9 in toluene	
Pm^{+3}		-1.10	22	HA<0.5	HS<0.9 in toluene	
Eu^{+3}		-0.44	22	HA<0.5	HS<0.9 in toluene	
	$M^{+3} + 3\overline{HS} = \overline{MS_3} + 3H^+$	2.17	22	μ = 0.025 HA	HS<0.3 in decanol	
	$M^{+3} + 6\overline{HS} = \overline{M(HS_2)_3}$	3.30	22	μ = 0.025 HA	HS<0.3 in decanol	64M
Tb^{+3}	$M^{+3} + 3\overline{(HS)_2} = \overline{M(HS_2)_3} + 3H^+$	0.56	22	HA<0.5	HS<0.9 in toluene	62B/57Pa
Tm^{+3}		2.20	22	HA<0.5	HS<0.9 in toluene	

METAL ION, M	REACTION	LOG K	TEMP.	CONDITIONS, AQ.	CONDITIONS, ORG.	REF.
Ligand: Cl^- , A^- (cont.)						
Tm^{+3}	$M^{+3} + 3\overline{HS} = \overline{MS_3} + 3H^+$	2.48	22	μ = 0.025 HA	HS<0.3 in decanol	64M
	$M^{+3} + 6\overline{HS} = \overline{M(HS_2)_3} + 3H^+$	5.30	22	μ = 0.025 HA	HS<0.3 in decanol	
Yb^{+3}	$M^{+3} + 3\overline{(HS)_2} = \overline{M(HS_2)_3} + 3H^+$	2.58	22	HA<0.5	HS<0.9 in toluene	62B/57Pa
Lu^{+3}		2.90	22	HA<0.5	HS<0.9 in toluene	
Th^{+4}	$M^{+4} + 4\overline{HS} = \overline{MS_4} + 4H^+$	4.66	22	μ = 0.25 HA	HS<0.5 in decanol	64M
UO_2^{++}	$M^{+2} + 2\overline{HS} = \overline{MS_2} + 2H^+$	1.93	22	μ = 0.25 HA	HS<0.3 in decanol	63Z
		1.62	22	μ = 0.25 HA	HS<0.5 in n-amyl alcohol	
	$M^{+2} + 4\overline{HS} = \overline{M(HS_2)_2} + 2H^+$	3.81	22	μ = 0.25 HA	HS<0.3 in decanol	63Z
		3.17	22	μ = 0.25 HA	HS 0.5 in n-amyl alcohol	
	$M^{+2} + 2\overline{(HS)_2} = \overline{M(HS_2)_2} + 2H^+$	4.53	?	HA ?	HS in kerosene	59B
		2.72	22	μ = 1.0 $(Na^+,H^+)A^-$	HS<0.1 in toluene	62P
		ΔG -5.14	25	μ = 0.75 HA 0.5	HS 0.15 in CCl_4	63M
		ΔH 0.04				
		ΔS 17.4				

METAL ION, M	REACTION	LOG K	TEMP.	CONDITIONS, AQ.	CONDITIONS, ORG.	REF.
Ligand: Cl^- , A^- (cont.)						
UO_2^{++}	$M^{+2} + 2A^- + m\overline{(HS)_2} = \overline{MA_2(HS)_{2m}}$ m > 1	ΔH 2.42	10-50	HA 1.0	HS 0.1 in kerosene	65S
		ΔH 3.42	10-50	HA 8.0	HS 0.1 in kerosene	
		ΔH 7.84	10-50	HA 4.0	HS 0.1 in kerosene	
	$M^{+2} + 2\overline{(HS)_2} + \overline{DBB'P} = \overline{MS\ HSH \cdot DBB'P}$	∿ 7.36	?	HA 0.3	HS 0.01, DBB'P <0.45 in cyclohexane	63Z
	$M^{+2} + 2\overline{(HS)_2} + \overline{TO'PO} = \overline{MS_3 \cdot HSH \cdot TO'PO}$	∿ 9.5	?	HA 0.3	HS 0.01, TO'PO<0.45 in cyclohexane	
Am^{+3}	$M^{+3} + 3\overline{HS} = \overline{MS_3} + 3H^+$	1.9	22	μ = 0.025 HA	HS<0.3 in decanol	64M
	$M^{+3} + 6\overline{HS} = \overline{M(HS_2)_3} + 3H^+$	2.53	22	μ = 0.025 HA	HS<0.3 in decanol	
	$M^{+3} + 3\overline{(HS)_2} = \overline{M(HS_2)_3} + 3H^+$	-1.67	?	HA<1.50	50% in toluene	66G
Cm^{+3}		-0.96	?	μ = 0.25 HA	HS in toluene	63M/59P, 61P
		-1.43	?	HA<1.50	50% in toluene	66G
Bk^{+3}		-0.4	?	HA 1.0	HS 50% in toluene	66S/66G
Cf^{+3}		-2.66	?	μ = 0.25 HA	HS in toluene	63M/59P, 61P
		-0.17	?	HA<1.5	HS 50% in toluene	66G
Fm^{+3}		+0.49	?	HA<1.5	HS 50% in toluene	
Md^{+3}		+0.8	?	HA 1.0	HS 50% in toluene	66S/66G
Co^{++}	$M^{+2} = 2\overline{(HS)_2} = \overline{M(HS_2)_2} + 2H^+$	-4.90	22	HA<2.0	HS 50% in toluene	62B/60K

METAL ION, M	REACTION	LOG K	TEMP.	CONDITIONS, AQ.	CONDITIONS, ORG.	REF.
Ligand: Cl^- , A^- (cont.)						
Cu^{++}	$M^{+2} + 2\overline{(HS)_2} = \overline{M(HS_2)_2} + 2H^+$	-3.52	21	HA<2.0	HS 50% in toluene	62B/60K
Ag^+	$M^+ + \overline{(HS)_2} = \overline{MHS_2} + H^+$	-1.84	21	HA<2.0	HS 50% in toluene	
Zn^{++}	$M^{+2} + 2\overline{(HS)_2} = \overline{M(HS_2)_2} + 2H^+$	-1.30	21	HA<2.0	HS 50% in toluene	
Cd^{++}		-3.75	21	HA<2.0	HS 50% in toluene	
Tl^+	$M^+ + \overline{(HS)_2} = \overline{MHS_2} + H^+$	-2.58	21	HA<2.0	HS 50% in toluene	
Ligand: ClO_4^- , A^-						
Ca^{++}	$M^{+2} + 2\overline{(HS)_2} = \overline{M(HS_2)_2} + 2H^+$	-2.6	25	HA ?	HS in toluene	62B/60Da
Sr^{++}		-3.8	25	HA ?	HS in toluene	
Sc^{+3}	$M^{+3} + 3\overline{(HS)_2} = \overline{M(HS_2)_3} + 3H^+$	7.8	?	$\mu = 1$ $(Na^+,N^+)A^-$	HS 0.015 in toluene	58P
		1.4	?	HA 0.1	HS<1.0 in toluene	62B/60D
Y^{+3}		0.8	?	$\mu = 1$ $(H^+,Na^+)A^-$	HS 0.15 in toluene	58P
La^{+3}		-3.3	?	$\mu = 1$ $(H^+,Na^+)A^-$	HS 0.15 in toluene	58P
		-2.4	?	HA 0.1	HS<1.0 in toluene	62B/60D
		-2.24	25	HA 1.0	HS 16% in toluene	66S/63P

METAL ION, M	REACTION	LOG K	TEMP.	CONDITIONS, AQ.	CONDITIONS, ORG.	REF.
Ligand: ClO_4^- , A^- (cont.)						
Ce^{+3}	$M^{+3} + 3\overline{(HS)_2} = \overline{M(HS_2)_3} + 3H^+$	-1.6	?	HA 0.1	HS<1.0 in toluene	62B/60D
		-1.77	25	HA 1.0	HS 16% in toluene	66S/63P
Pr^{+3}		-1.6	?	HA 0.1	HS<1.0 in toluene	62B/60D
		-1.46	25	HA 1.0	HS 16% in toluene	66S/63P
Nd^{+3}		-1.32	25	HA 1.0	HS 16% in toluene	
Pm^{+3}		-2	?	$\mu = 1\ (H^+,Na^+)A^-$	HS<0.2 (?) in toluene	61P
		-0.83	?	$\mu = 1\ (H^+,Na^+)A^-$	HS 0.15 in toluene	62B/58P
		-1.10	?	HA 0.1	HS<1.0 in toluene	62B/60D
		-0.98	25	HA 1.0	HS 16% in toluene	66S/63P
Sm^{+3}		-0.8	?	HA 0.1	HS<1.0 in toluene	62B/60D
		-0.47	25	$\mu = 1.00\ (H^+,Na^+)A^-$	HS 0.05 in toluene	66K
Eu^{+3}		-0.50	?	HA 0.1	HS<1.0 in toluene	62B/60D
		-0.35	25	$\mu = 1.00\ (H^+,Na^+)A^-$	HS 0.05 in toluene	66K
Gd^{+3}		0.20	?	HA 0.1	HS<1.0 in toluene	62B/60D
		-0.06	25	HA 1.0	HS 16% in toluene	66S/63P

METAL ION, M	REACTION	LOG K	TEMP.	CONDITIONS, AQ.	CONDITIONS, ORG.	REF.
Ligand: ClO_4^- , A^- (cont.)						
Tb^{+3}	$M^{+3} + 3\overline{(HS)_2} = \overline{M(HS_2)_3} + 3H^+$	0.50	?	HA 0.1	HS<1.0 in toluene	62B/60D
		0.69	25	μ = 1.00 $(H^+,Na^+)A^-$	HS 0.05 in toluene	66K
		0.63	25	HA 1.0	HS 16% in toluene	66S/63P
Dy^{+3}		1.06	25	μ = 1.00 $(H^+,Na^+)A^-$	HS 0.05 in toluene	66K
Ho^{+3}		1.40	?	HA 0.1	HS<1.0 in toluene	62B/60D
		1.45	25	μ = 1.00 $(H^+,Na^+)A^-$	HS 0.05 in toluene	66K
		1.24	25	HA 1.0	HS 16% in toluene	66S/63P
Er^{+3}		1.93	25	μ = 1.00 $(H^+,Na^+)A^-$	HS 0.05 in toluene	66K
		1.59	25	HA 1.0	HS 16% in toluene	66S/63P
Tm^{+3}		2.27	?	μ = 1 $(H^+,Na^+)A^-$	HS 0.15 in toluene	62B/58P
		2.50	?	HA 0.1	HS<1.0 in dibutyl ether	62B/60D
		2.50	25	μ = 1.00 $(H^+,Na^+)A^-$	HS 0.05 in toluene	66K
		1.99	25	HA 1.0	HS 16% in toluene	66S/63P
Yb^{+3}		2.97	25	μ = 1.00 $(H^+,Na^+)A^-$	HS 0.05 in toluene	66K
		2.48	25	HA 1.0	HS 16% in toluene	66S/63P
Lu^{+3}		4.1	?	HA 0.1	HS<1.0 in toluene	
		3.14	25	μ = 1.00 $(H^+,Na^+)A^-$	HS 0.05 in toluene	66K

METAL ION, M	REACTION	LOG K	TEMP.	CONDITIONS, AQ.	CONDITIONS, ORG.	REF.
Ligand: ClO_4^- , A^- (cont.)						
Lu^{+3}	$M^{+3} + 3\overline{(HS)_2} = \overline{M(HS_2)_3} + 3H^+$	2.75	25	HA 1.0	HS 16% in toluene	66S/63P
Ac^{+3}		-4	?	$\mu = 1\ (H^+,Na^+)A^-$	HS 0.15 in toluene	58P
UO_2^{++}	$M^{+2} = 2S^- = \overline{MS_2}$	13.53	20	$\mu = 0.1\ (H^+,Na^+)A^-$	HS 0.1 in octane	66U
	$M^{+2} + 2S^- + 2HS = \overline{MS_2 \cdot 2HS}$	29.77	20	$\mu = 0.1\ (H^+,Na^+)A^-$	HS 0.1 in octane	
	$M^{+2} + 2\overline{HS} = \overline{MS_2} + 2H^+$	3.85	22	HA 1.0	HS<0.02 in decane	64G
		3.85	20	$\mu = 0.1\ (H^+,Na^+)A^-$	$HS<10^{-4}$ in octane	66U
	$M^{+2} + 2\overline{(HS)_2} = \overline{M(HS_2)_2} + 2H^+$	4.60	25	$\mu = 2\ (H^+,Na^+)A^-$	HS<1 in hexane	58B
		4.45	25	$\mu = 2\ (H^+,Na^+)A^-$	HS 0.005 in hexane	62B/60Ha
		2.98	22	$\mu = 1.0\ (H^+,Na^+)A^-$	HS<0.1 in toluene	62P
		ΔG -5.36	25	$\mu = 0.75$ HA 0.5	HS 0.15 in CCl_4	63M
		ΔS 0.50				
		ΔS 19.7				
	$M^{+2} + 2\overline{(HS)_2} = \overline{MS_2 \cdot 2HS} + 2H^+$	4.27	20	$\mu = 0.1\ (H^+,Na^+)A^-$	HS<0.1 in octane	66U
NpO_2^{++}	$M^{+2} + 2\overline{(HS)_2} = \overline{M(HS_2)_2} + 2H^+$	4.36	21	HA<0.52	HS<5.0% in toluene	62B/61N
Pu^{+4}	$M^+ + 4\overline{HS} = \overline{MS_4} + 4H^+$	6.19	22	HA 1.0	HS 0.01 in decane	64G

METAL ION, M	REACTION	LOG K	TEMP.	CONDITIONS, AQ.	CONDITIONS, ORG.	REF.
Ligand: ClO_4^- , A^- (cont.)						
Am^{+3}	$M^{+3} + 3\overline{(HS)_2} = \overline{M(HS_2)_3} + 3H^+$	-1.53	?	μ = 1.0 $(H^+,Na^+)A^-$	HS 0.15 in toluene	62B/58P
		-1.53	?	μ = 1 $(H^+,Na^+)A^-$	HS 0.15 in toluene	66G/58P
Cm^{+3}		-2.5	?	μ = 1 $(H^+,Na^+)A^-$	HS 0.2 (?) in toluene	61P
		-1.40	?	μ = 1 $(H^+,Na^+)A^-$	HS 0.15 in toluene	66G/58P
Cf^{+3}		0.0	?	μ = 1 $(H^+,Na^+)A^-$	HS 0.2 (?) in toluene	66G/61P
Fe^{+3}		6.55	25	μ = 2 $(H^+,Na^+)A^-$	HS<0.5 in n-octane	60B
	$\overline{0.8\ MS_6H_3} + M^{+3} + 0.6A^-$	2.39	25	μ = 2 $(H^+,Na^+)A^-$	HS<0.1 in n-octane	
	$= \overline{0.15\ [M_3S_8A]_4} + 2.4H^+$			$[M^{+3}]/[HS] < 0.375$		
	$\overline{0.25[M_3S_8A]_4} + 2M^{+3} + 6H_2O$	-4.92	25	μ = 2 $(Na^+,H^+)A^-$	HS 0.1 in n-octane	
	$= \overline{M_5S_8A(OH)_6} + 6H^+$			$[M^{+3}]/[HS]$ 0.625		
	$M^{+3} + 3\overline{(HS)_2} = \overline{M(HS_2)_3} + 3H^+$	3.6	25	μ = 2 $(H^+,Na^+)A^-$	HS<0.3 in toluene	62B/60B
Hf^{+4}	$M^{+4} + A^- + 5\overline{HS} = \overline{MAS(HS_2)_2} + 3H^+$	38.5	19	μ = 2.00 $(H^+,Na^+)A^-$	HS 2.10^{-4} in octane	67N

REFERENCES

55B Blake, C. A., Brown, K. B., Coleman, C. F., White, J. C. and Schmitt, J. M., USAEC Report ORNL - 1903 (1955)

57P Peppard, D. F., Mason, G. W. and Moline, S. W., J. Inorg. Nucl. Chem., 5, 141 (1957)

57Pa Peppard, D. F., Mason, G. W., Maier, J. L. and Driscoll, W. J., J. Inorg. Nucl. Chem., 4, 334 (1957)

58B Baes, C. F., Jr., Zingaro, R. A. and Coleman, C. F., J. Phys. Chem., 62, 129 (1958)

58Ba Blake, C. A., Baes, C. F. Jr., Brown, K. B., Coleman, C. F. and White, S. C., Proc. Conf. Peaseful Uses At.En. 28, 289 (1958)

58P Peppard, D. F., Mason, G. W., Driscoll, W. J. and Sironen, R. J., J. Inorg. Nucl. Chem., 7, 276 (1958)

58Pa Peppard, D. F., Mason, G. W. and Moline, S. W., J. Inorg. Nucl. Chem., 5, 141 (1958)

59H Hardy, C. J. and Greenfield, B. F., unpublished work (1959)

59B Blake, C. A., Horner, D. E. and Schmitt, J. M., USAEC Report ORNL - 2259 (1959)

59P Peppard, D. F., Mason, G. W., Driscoll, W. J. and McCarty, S., J. Inorg. Nucl. Chem., 12, 141 (1959)

60B Baes, C. F., Jr., and Baker, H. T., J. Phys. Chem., 64, 89 (1960)

60D Duyckaerts, G., Dreze, Ph. and Simon, A., J. Inorg. Nucl. Chem., 13, 332 (1960)

60Da Dyrssen, D. and Liem, D. H., Acta Chem. Scand., 14, 1100 (1960)

60H Hardy, C. J., Greenfield, B. F. and Scargill, D., British Report AERE/R - 3316 (1960)

60Ha Hardy, C. J., British Report AERE/R - 3267 (1960)

60K Kimura, K., Bull. Chem. Soc. Japan, 33, 1038 (1960); 34, 63 (1961)

60L Lamb, E., USAEC Report CF-60-6-132 (1960)

60P Peppard, D. F., Mason, G. W. and McCarty, S., J. Inorg. Nucl. Chem., 13, 138 (1960)

61H Hardy, C. J., Greenfield, B. F. and Scargill, D., J. Chem. Soc., 174 (1961)

61Ha Hardy, C. J., J. Inorg. Nucl. Chem., 21, 348 (1961)

61N Nakamura, E., Bull. Chem. Soc. Japan, 34, 402 (1961)

61P Peppard, D. F. Mason, G. W. and Hucher, I., J. Inorg. Nucl. Chem., 18, 245 (1961)

62B Baes, C. F., Jr., J. Inorg. Nucl. Chem., 24, 707 (1962)

62Ba Baker, H. T. and Baes, C. F., Jr., J. Inorg. Nucl. Chem., 24, 1277 (1962)

62P Peppard, D. F., Mason, G. W., Hucher, I. and Brandao, F. A., J. Inorg. Nucl. Chem., 24, 1387 (1962)

62Pa Peppard, D. F., Mason, G. W., McCarty, S. and Johnson, F. D., J. Inorg. Nucl. Chem., 24, 321 (1962)

62S Sato, T., J. Inorg. Nucl. Chem., 24, 699 (1962)

63B Baes, C. F., Jr., Nucl. Sci. & Eng., 16, 405 (1963)

63I Ihle, H., Michael, H. and Murrenhoff, A., J. Inorg. Nucl. Chem., 25, 734 (1963); German Report JUL-74-PC (1962)

63M Mitsuru Asano, Okajima, Y. and Tomota Nishi, Tech. Rept. Eng. Res. Inst. Kyoto Univ, (1963)

63P Pierce, T. B. and Peck, P. F., Analyst., 88, 217 (1963)

63S Sato, T., J. Inorg. Nucl. Chem., 25, 109 (1963)

63U Ulyanov, V. S. and Sviridova, R. A., Radiokhimiya, 5, 419 (1963)

63Z Zangen, M., J. Inorg. Nucl. Chem., 25, 581 (1962)

64G Gureev, E. S., Kosyakov, V. N. and Yakovlev, G. N., Radiokhimiya, 6, 655 (1964)

64M Mason, G. W., Lewey, S. and Peppard, D. F., J. Inorg. Nucl. Chem., 26, 2271 (1964)

64S Sato, T., J. Inorg. Nucl. Chem., 26, 311 (1964)

65B Baram, I. I., Zh. Neorg. Khim., 10, 2184 (1965)

65D Deptula, C. and Minc, S., Nukleonika, 10, 421 (1965)

65S Sato, T., J. Inorg. Nucl. Chem., 27, 1853 (1965)

65Sa Sato, T., J. Inorg. Nucl. Chem., 27, 1395 (1965)

66G Gavrilov, K. A., Gwozdz, E., Stary, J. and Wang Tung-Seng, Talanta, 13, 471 (1966)

66K Kolarik, Z. and Pankova, H., J. Inorg. Nucl. Chem., 28, 2325 (1966)

66S Stary, J., Talanta, 13, 421 (1966)

66Sa Szabo, E. and Szabon J., Acta Chim. Acad. Sci. Hungaricae, 48, 299 (1966)

66U Ul'yanov, V. S., Sviridova, R.A. and Laskorin, B. N., Radiokhimiya, 8, 416 (1966)

66Z Zangen, M., J. Inorg. Nucl. Chem., 28, 1963 (1966)

67K Kolarik, Z., Collec. Czech. Chem. Commun., 32, 311 (1967)

67N Navratil, O., in "Solvent Extraction Chemistry", Dyrssen, D., Liljenzin, J. O., Rydberg, J. (eds.) North-Holland, Amsterdam, p. 256 (1967)

67S Sato , T., J. Inorg. Nucl. Chem., 29, 555 (1967)

68S Asano, M., and Kobsaka, N., Tech. Rep. Eng. Res. Inst. Kyoto Univ., No. 140 (1968)

69S Smilov, V. C., Lanin, V.P., Radiokhimiya, 11, 415 (1969)

METAL ION, M	REACTION	LOG K	TEMP.	CONDITIONS, AQ.	CONDITIONS, ORG.	REF.
Extractant: Diisopentyl phosphoric acid (HDiAP), HS						
Dissociation:	$HS = H^+ + S^-$	-1.14	?	$\mu = 1\ (Na^+,H^+)NO_3^-$, 10^{-4} S	HS<0.05 in toluene	67K
Distribution:	$HS = \overline{HS}$	1.11	?			
Dimerization:	$2\overline{HS} = \overline{H_2S_2}$	3.84	?			
Reaction with other organophosphorus compounds						
UO_2^{++}	$\overline{MS_2} + \overline{TBP} = \overline{MS_2 \cdot TBP}$	4.34	10	$\mu = 2\ HNO_3$	HS $2 \cdot 10^{-3}$, TBP $6.3 \cdot 10^{-3}$ in xylene	61S
		3.94	60			
		4.14	22	$\mu = 2\ HNO_3$	HS $3 \cdot 10^{-3}$, TBP<0.1 in benzene	62Sa
	$\overline{MS_2} + \overline{DAM'P} = \overline{MS_2 \cdot DAM'P}$	4.0	?	$\mu = 2\ HNO_3$	DAM'P<0.01 in benzene	62S
	$\overline{MS_2} + \overline{TB'PO} = \overline{MS_2 \cdot TB'PO}$	5.1	?	$\mu = 2\ HNO_3$	TB'PO<0.01 in benzene	
Pu^{+4}	$\overline{MS_4} + \overline{TBP} = \overline{MS_4 \cdot TBP}$	2.74	10-60	$\mu = 2\ HNO_3$	HS $2 \cdot 10^{-4}$, TBP 0.02 in xylene	61S
Ligand: NO_3^-, A^-						
UO_2^{++}	$M^{+2} + 2\overline{HS} = \overline{MS_2} + 2H^+$	4.93	10	HA 2	HS $2 \cdot 10^{-3}$ in xylene	61S
		3.88	60			
		4.34	22	HA 2.0	HS 10^{-3} in benzene	62S

METAL ION, M	REACTION	LOG K	TEMP.	CONDITIONS, AQ.	CONDITIONS, ORG.	REF.
Ligand: ClO_4^- , A^-						
Hf^{+4}	$M^{+4} + A^- + \overline{5HS} = \overline{MAS(HS_2)_2} + 3H^+$	28.1	19	$\mu = 2.00\ (H^+,Na^+)A^-$	HS 3.10^{-3} in toluene	67N

References

61S Shevchenko, V. B., Fedorov, I. A. and Smelov, V. S., Radiokhimiya, 3, 256 (1961)

62S Shevchenko, V. B., Smelov, V. S. and Strakhova, A. V., Zh. Neorg. Khim., 7, 1736 (1962)

62Sa Shevchenko, V. B., Smelov, V. S. and Strakhova, A. V., Ekst. Teor. Prim. App. Sb. Stat. No. 2, p. 179 (1962)

67K Kolarik, Z., Coll. Czech. Chem. Commun., 32, 311 (1967)

67N Navratil, O., in "Solvent Extraction Chemistry", Eds. Dyrssen, D., Liljenzin, J.O., Rydberg, J., North-Holland, Amsterdam, p. 256 (1967)

METAL ION, M	REACTION	LOG K	TEMP.	CONDITIONS, AQ.	CONDITIONS, ORG.	REF.
Extractant: Dioctylphosphoric acid (HDOP), HS						
Dissociation						
	$HS = H^+ + S^-$	-1.2	20	$\mu = 0.1\ (Na^+, H^+)\ ClO_4^-$, pH 3	HS<0.1 in n-octane	65U
		-2.9	25		HS in ethanol 75%	55B
		-1.3	20	$\mu = 0.1\ (Na^+, H^+)\ ClO_4^-$	HS<0.1 in n-octane	67K
Distribution						
	$HS = \overline{HS}$	3.98	20	$\mu = 0.1\ (Na^+, H^+)\ ClO_4^-$	HS<0.1 in n-octane	65U
		3.80	20	$\mu = 0.1\ (Na^+, H^+)\ ClO_4^-$	HS<0.1 in n-octane	67K
Dimerization						
	$2\overline{HS} = \overline{(HS)_2}$	3.50	20	$\mu = 0.1\ (Na^+, H^+)\ ClO_4^-$, pH3.5	HS<0.1 in octane	65U, 67K
Ligand HS						
Nd^{+3}	$MS_3 + 1.5\overline{(HS)_2} = \overline{MS_3(HS)_3}$	0.04	25		HS<1 in benzene	63K
Yb^{+3}		-0.36				
Ligand NO_3^-, A^-						
Be^{+2}	$M^{+2} + 2\overline{(HS)_2} = \overline{M(HS_2)_2} + 2H^+$	1.71	?	HA 0.25	HS 0.5 in toluene	61H
Y^{+3}	$M^{+3} + 3\overline{(HS)_2} = \overline{M(HS_2)_3} + 3H^+$	3.46	25	$\mu = 1.00\ (H^+, Na^+) A^-$	HS 0.05 in toluene	66K

METAL ION, M	REACTION	LOG K	TEMP.	CONDITIONS, AQ.	CONDITIONS, ORG.	REF.
Ligand: NO_3^- , A^- (cont.)						
La^{+3}	$M^{+3} + 3\overline{(HS)_2} = \overline{M(HS_2)_3} + 3H^+$	-0.61	25	$\mu = 1.00\ (H^+,Na^+)A^-$	HS 0.05 in toluene	66K
Ce^{+3}		-0.03				
Pm^{+3}		0.47				
Sm^{+3}		1.00				
Eu^{+3}	$M^{+3} + 2\overline{(HS)_2} = \overline{MS_2(HS_2)} + 3H^+$	0.42			HS 0.05 in MiBK	67Ka
	$M^{+3} + 2.5\overline{(HS)_2} = \overline{MS(HS_2)_2} + 3H^+$	2.81			HS 0.05 in 2,2,4-trimethyl-pentane	
		2.98			HS 0.05 in n-dodecane	
		2.98			HS 0.05 in n-heptane	
		2.55			HS 0.05 in cyclohexane	
		0.89			HS 0.05 in nitrobenzene	
	$M^{+3} + 3\overline{(HS)_2} = \overline{M(HS_2)_3} + 3H^+$	2.14			HS 0.05 in n-butylether	
		2.06			HS 0.05 in CCl_4	
		1.48			HS 0.05 in p-xylene	
		1.43			HS 0.05 in o-xylene	
		1.45			HS 0.05 in bromoform	
		1.20			HS 0.05 in chlorobenzene	

METAL ION, M	REACTION	LOG K	TEMP.	CONDITIONS, AQ.	CONDITIONS, ORG.	REF.
Ligand: NO_3^- , A^- (cont.)						
Eu^{+3}	$M^{+3} + 3\overline{(HS)_2} = \overline{M(HS_2)_3} + 3H^+$	1.13	25	$\mu = 1.00\ (H^+, Na^+)A^-$	HS 0.05 in benzene	67Ka
		0.92			HS 0.05 in methylene chloride	
		0.25			HS 0.05 in chloroform	
Tb^{+3}	$M^{+3} + 2\overline{(HS)_2} = \overline{MS_2(HS_2)} + 3H^+$	1.42			HS 0.05 in MiBK	67Ka
	$M^{+3} + 2.5\overline{(HS)_2} = \overline{MS(HS_2)_2} + 3H^+$	3.74			HS 0.05 in 2,2,4-trimethyl-pentane	
		3.84			HS 0.05 in n-dodecane	
		3.88			HS 0.05 in n-heptane	
		3.46			HS 0.05 in cyclohexane	
		1.90			HS 0.05 in nitrobenzene	
	$M^{+3} + 3\overline{(HS)_2} = \overline{M(HS_2)_3} + 3H^+$	3.10			HS 0.05 in n-butylether	
		3.08			HS 0.05 in carbon tetra-chloride	
		2.60			HS 0.05 in p-xylene	
		2.55			HS 0.05 in o-xylene	
		2.49			HS 0.05 in bromoform	

METAL ION, M	REACTION	LOG K	TEMP.	CONDITIONS, AQ.	CONDITIONS, ORG.	REF.
Ligand: NO_3^- , A^- (cont.)						
Tb^{+3}	$M^{+3} + 3\overline{(HS)_2} = \overline{M(HS_2)_3} + 3H^+$	2.37	25	$\mu = 1.00\ (H^+,Na^+)A^-$	HS 0.05 in toluene	67Ka
		2.25			HS 0.05 in chlorobenzene	
		2.11			HS 0.05 in benzene	
		2.00			HS 0.05 in methylene chloride	
		1.25			HS 0.05 in chloroform	
Dy^{+3}		2.98	25		HS 0.05 in toluene	66K
Ho^{+3}		3.32				
Er^{+3}		3.74				
Tm^{+3}		4.34				
Yb^{+3}		4.76				
Lu^{+3}		5.08				
Ligand: ClO_4^- , A^-						
La^{+3}	$M^{+3} + 3\overline{(HS)_2} = \overline{M(HS_2)_3} + 3H^+$	-0.25	25	$\mu = 1.00\ (H^+,Na^+)A^-$	HS 0.05 in toluene	66K
Ce^{+3}		0.33				
Pm^{+3}		0.92				
Sm^{+3}		1.48				

METAL ION, M	REACTION	LOG K	TEMP.	CONDITIONS, AQ.	CONDITIONS, ORG.	REF.
Ligand: ClO_4^- , A^- (cont.)						
Eu^{+3}	$M^{+3} + 3\overline{(HS)_2} = \overline{M(HS_2)_3} + 3H^+$	1.68	25	μ = 1.00 $(H^+,Na^+)A^-$	HS 0.05 in toluene	66K
Tb^{+3}		2.70				
Dy^{+3}		3.16				
Ho^{+3}		3.52				
Er^{+3}		3.92				
Tm^{+3}		4.57				
Yb^{+3}		4.96				
Lu^{+3}		5.28				
Hf^{+4}	$M^{+4} + A^- + 3HS = \overline{MAS_3} + 3H^+$	25.5	19	μ = 2.00 (H^+,Na^+) A^-	HS 4.10^{-5} in octane	67N

References

55B Blake, C. A., Brown, K. B., Coleman, C. F., Horner, D. E. and Schmitt, J. M. USAEC Report ORNL 1903 (1955)

61H Hardy, C. J., Greenfeld, B. F. and Scargill, D., J. Chem. Soc. 174 (1961)

63K Kriss, E. E., Zh. Neorg. Khim., 8, 1505 (1963)

65U Ul'yanov, V. S. and Sviridova, R. A., Radiokhimiya, 7, 538 (1965)

66K Kolarik, Z. and Pankova, H., J. Inorg. Nucl. Chem., 28, 2325 (1966)

67K Kolarik, Z., Coll. Czech. Chem. Commun., 32, 311 (1962)

67Ka Kolarik, Z., in "Solvent Extraction Chemistry", Eds. Dyrssen, D., Liljenzin, J. O., Rydberg, S., North-Holland, Amsterdam, p. 250 (1967)

67N Navratil, O., in "Solvent Extraction Chemistry", Eds. Dyrssen, D., Liljenzin, J. O., Rydberg, J., North-Holland, Amsterdam, p. 256 (1967)

METAL ION, M	REACTION	LOG K	TEMP.	CONDITIONS, AQ.	CONDITIONS, ORG.	REF.
Extractant: Mono-Butylphosphoric acid (H_2MBP) , H_2S						
Dissociation						
	$H_2S = H^+ + HS^-$	-1.89	25	H_2S in H_2O	——	43K
	$\overline{H_2S} = \overline{H^+} + \overline{HS^-}$	-3.90	?	——	HS 0.05 in ethanol 98%	65D
	$HS^- = H + S^=$	-6.84	25	H_2S in H_2O	——	43K
Ligand: NO_3^- , A^-						
Zr^{+4}	$M(HS)_3 + \overline{H_2S} = \overline{M(HS)_4} + H^+$	1.94	20	LiA 3.0, HA 3.0	H_2S<0.075 in kerosene	62S
Nb^{+5}	$M(OH)_3 \cdot (HS)_3^- + 2H_2S + H^+ = \overline{Nb(HS)_5} + 3H_2O$	1.96	20	LiA 3.0, HA 3.0	H_2S<0.1 in kerosene	
UO_2^{++}	$M^{+2} + \overline{2H_2S} = \overline{M(HS)_2} + 2H^+$	2.36	20	LiA 2.0, HA 2.0	H_2S<0.1 in hydrogenated kerosene	62Sa
Ligand: $SO_4^=$, $A^=$						
H^+	$\overline{2R_3N \cdot H_2S} + 2H^+ + A^= = \overline{(R_3NH)_2A} + \overline{2H_2S}$	2.42	?	H_2A 2.08	HS 0.1, R_3N 0.1 in benzene	65D

R_3N - trioctylamine

References

43K Kumler, W. D. and Eiler, J., J. Am. Chem. Soc., 65, 2355 (1943)

62S Shevchenko, V. B. and Smelov, V. S., Ekst. Teor. Prim. App. Sb. Sta. No. 2, p. 257 (1962)

62Sa Shevchenko, V. B. and Smelov, V. S., Ekst. Teor, Prim. App. Sb. Sta. No. 2, p. 219 (1962)

65D Deptula, C. and Minc, S., Nukleonika, 10, 421 (1965)

METAL ION, M	REACTION	LOG K	TEMP.	CONDITIONS, AQ.	CONDITIONS, ORG.	REF.
Extractant: Mono (2-ethyl-hexyl) phosphoric acid, (H_2MEHP) , H_2S						
Dissociation						
	$\overline{H_2S} = \overline{H^+} + \overline{HS^-}$	-3.1	25		HS in 75% ethanol	55B
	$\overline{HS^-} = \overline{H^+} + \overline{S^=}$	-8.4	25			
Ligand: ClO_4^- , A^-						
Sc^{+3}	$M^{+3} + \overline{(H_2S)_6} = \overline{M[H(HS)_2]_3} + 3H^+$	6.5	?	HA 2	$H_2S<0.7$ in toluene	58P
Y^{+3}		1.4	?	HA 0.2	$H_2S<0.7$ in toluene	
La^{+3}		1	?	HA 0.05	$H_2S<0.7$ in toluene	
Ac^{+3}		1	?	HA 0.05	$H_2S<0.7$ in toluene	

References

55B Blake, C. A., Brown, K. B., Coleman C. F. and Schmitt, J. M., USAEC Report ORNL - 1903 (1955)

58P Peppard, D. F., Mason, G. W., Driscoll, W. J. and Sironen, R. J., J. Inorg. Nucl. Chem., 7, 276 (1958)

METAL ION, M	REACTION	LOG K	TEMP.	CONDITIONS, AQ.	CONDITIONS, ORG.	REF.
Extractant: Di-p-chlorophenyl Phosphoric acid, (H)DpCl-PhP, HS						
Distribution:						
	$HS = \overline{HS}$	0.06	rt	$HClO_4$ 0.1	HS<0.01 in benzene	66K
		-0.42	rt	$HClO_4$ 0.1	HS<0.01 in CCl_4	66K
		0.80	20	$\mu = 1$ $(H^+, Na^+)ClO_4^-$	HS<0.01 in $CHCl_3$	63K,66Ka
		-1.05	rt	$HClO_4$ 0.1	HS<0.01 in kerosene	66K
		2.05	rt	$HClO_4$ 0.1	HS<0.01 in MiBK	66K
		2.64	rt	$HClO_4$ 0.1	HS<0.01 in n-octanol	66K
		1.63	rt	$HClO_4$ 0.1	HS<0.01 in i-propylether	66K
Dissociation:						
	$HS = H^+ + S^-$	-0.2	20	$\mu = 1$ $(H^+, Na^+)ClO_4^-$	HS<0.01 in $CHCl_3$	63K,66Ka
Dimerization:						
	$2\overline{HS} = \overline{(HS)_2}$	4.20	rt	$HClO_4$ 0.1	HS<0.01 in benzene	66K
		4.31	rt	$HClO_4$ 0.1	HS<0.01 in CCl_4	66K
		3.44	20	$\mu = 1$ $(H^+, Na^+)ClO_4^-$	HS<0.01 in $CHCl_3$	63K.66Ka
		4.81	rt	$HClO_4$ 0.1	HS<0.01 in kerosene	66K
		0.88	rt	$HClO_4$ 0.1	HS<0.01 in MiBK	66K
		very low	rt	$HClO_4$ 0.1	HS<0.01 in n-octanol	66K
		1.31	rt	$HClO_4$ 0.1	HS<0.01 in i-propylether	66K

METAL ION, M	REACTION	LOG K	TEMP.	CONDITIONS, AQ.	CONDITIONS, ORG.	REF.
Reactions with diluents and other solvents						
	$\overline{(HS)_2} + \overline{C_6H_6} = \overline{(HS)_2 \cdot C_6H_6}$	0.54	rt	$HClO_4$ 0.1	HS<3.5 x 10^{-4}, C_6H_6<0.01 in CCl_4	66K
	$\overline{HS} + 1.5\ \overline{BzOH} = \overline{HS \cdot 1.5BzOH}$	2.86	rt	$HClO_4$ 0.1	HS<3.5 x 10^{-4}, BzOH 0.01-0.1 in CCl_4	66K
	$\overline{HS} + \overline{BuAt} = \overline{HS \cdot BuAt}$	1.26	rt	$HClO_4$ 0.1	HS<3.5 x 10^{-4}, BuAt 0.01-1.0 in CCl_4	66K
	$\overline{(HS)_2} + \overline{CCl_4} = \overline{(HS)_2 \cdot CCl_4}$	-0.34	rt	$HClO_4$ 0.1	HS + CCl_4	66K
	$\overline{HS} = \overline{CHCl_3} = \overline{HS \cdot CHCl_3}$	0.14	rt	$HClO_4$ 0.1	HS<3.5 x 10^{-4}, $CHCl_3$ 0.1-5.0 in CCl_4	66K
	$\overline{(HS)_2} + \overline{CHCl_3} = \overline{(HS)_2 \cdot CHCl_3}$	0.79	rt	$HClO_4$ 0.1	HS<3.5 x 10^{-4}, $CHCl_3$ 0.1-5.0 in CCl_4	66K
	$\overline{HS} + \overline{cHxK} = \overline{HS \cdot cHxK}$	1.76	rt	$HClO_4$ 0.1	HS<3.5 x 10^{-4}, cHxK 0.02-1.0 in CCl_4	66K
	$\overline{HS} + 1.5\ \overline{cHxOH} = \overline{HS \cdot 1.5cHxOH}$	3.46	rt	$HClO_4$ 0.1	HS<3.5 x 10^{-4}, cHxOH 0.1-1.0 in CCl_4	66K
	$\overline{HS} + \overline{DBE} = \overline{HS \cdot DBE}$	1.24	rt	$HClO_4$ 0.1	HS<3.5 x 10^{-4}, DBE 0.01-1.0 in CCl_4	66K
	$\overline{HS} + \overline{DiPE} = \overline{HS \cdot DiPE}$	1.48	rt	$HClO_4$ 0.1	HS<3.5 x 10^{-4}, DiPE 0.01-4.0 in CCl_4	66K
	$\overline{(HS)_2} + \overline{DiPE} = \overline{(HS)_2DiPE}$	0.98	rt	$HClO_4$ 0.1	HS<3.5 x 10^{-4}, DiPE 0.01-4.0 in CCl_4	66K
	$\overline{HS} + \overline{D(pCl\text{-}Ph)Ph'P} = \overline{HS \cdot D(pCl\text{-}Ph)Ph'P}$	2.40	rt	$HClO_4$ 0.1	HS<3.5 x 10^{-4}<<D(pCl-Ph)Ph'P in CCl_4	67K,68K
	$\overline{HS} + \overline{DPhPh'P} = \overline{HS \cdot DPhPh'P}$	2.60	rt	$HClO_4$ 0.1	HS<3.5 x 10^{-4}<<DPhPh'P in CCl_4	67K,68K

METAL ION, M	REACTION	LOG K	TEMP.	CONDITIONS, AQ.	CONDITIONS, ORG.	REF.
Reactions with diluents and other solvents						
	$\overline{HS} + \overline{EDB'P} = \overline{HS \cdot EDB'P}$	4.28	rt	$HClO_4$ 0.1	$HS < 3.5 \times 10^{-4} \ll EDB'P$ in CCl_4	67K,68K
	$\overline{HS} + 1.5\ \overline{MiBC} = \overline{MS \cdot 1.5\ MiBC}$	3.46	rt	$HClO_4$ 0.1	$HS < 3.5 \times 10^{-4} \ll MiBC$ in CCl_4	66K
	$\overline{HS} + \overline{MiBK} = \overline{HS \cdot MiBK}$	1.46	rt	$HClO_4$ 0.1	$HS < 3.5 \times 10^{-4}$, MiBK 0.01-4.0 in CCl_4	
	$\overline{(HS)_2} + \overline{MiBK} = \overline{(HS)_2 \cdot MiBK}$	1.36	rt	$HClO_4$ 0.1	$HS < 3.5 \times 10^{-4}$, MiBK 0.01-4.0 in CCl_4	66K
	$\overline{HS} + 1.5\ \overline{OcOH} = \overline{MS \cdot 1.5\ OcOH}$	3.60	rt	$HClO_4$ 0.1	$HS < 3.5 \times 10^{-4} \ll OcOH$ in CCl_4	66K
	$\overline{HS} + \overline{PhDPh'P} = \overline{HS \cdot PhDPh'P}$	3.18	rt	$HClO_4$ 0.1	$HS < 3.5 \times 10^{-4} \ll PhDPh'P$ in CCl_4	67K,68K
	$\overline{HS} + \overline{TBP} = \overline{HS \cdot TBP}$	3.28	rt	$HClO_4$ 0.1	$HS < 0.35 \times 10^{-4} \ll TBP$ in CCl_4	68K
	$\overline{HS} + \overline{TB'PO} = \overline{HS \cdot TB'PO}$	4.98	rt	$HClO_4$ 0.1	$HS < 3.5 \times 10^{-4} \ll TB'PO$ in CCl_4	67K,68K
	$\overline{HS} + \overline{TOP} = \overline{HS \cdot TOP}$	3.38	rt	$HClO_4$ 0.1	$HS < 3.5 \times 10^{-4} \ll TOP$ in CCl_4	68K
	$\overline{HS} + \overline{TO'PO} = \overline{HS \cdot TO'PO}$	4.98	rt	$HClO_4$ 0.1	$HS < 3.5 \times 10^{-4} \ll TO'PO$ in CCl_4	68K
	$\overline{HS} + \overline{TPhP} = \overline{HS \cdot TPhP}$	2.60	rt	$HClO_4$ 0.1	$HS < 3.5 \times 10^{-4} \ll TPhP$ in CCl_4	68K
	$\overline{HS} + \overline{TPh'PO} = \overline{HS \cdot TPh'PO}$	3.72	rt	$HClO_4$ 0.1	$HS < 3.5 \times 10^{-4} \ll TPh'PO$ in CCl_4	68K

METAL ION, M	REACTION	LOG K	TEMP.	CONDITIONS, AQ.	CONDITIONS, ORG.	REF.
Ligand: ClO_4^- , A^-						
Y^{+3}	$M^{+3} + 3S^- + 2\overline{HS} = \overline{MS_3(HS)_2}$	18.00	20	$\mu = 1\ (H^+,Na^+)ClO_4^-$	HS<0.01 in $CHCl_3$	66Ka
	$M^{+3} + 2.5\overline{(HS)_2} = \overline{MS_3(HS)_2} + 3H^+$	6.40	20	$\mu = 1\ (H^+,Na^+)ClO_4^-$	HS<0.01 in $CHCl_3$	66Ka
Ce^{+3}	$M^{+3} + 3S^- + 2\overline{HS} = \overline{MS_3(HS)_2}$	16.50	20	$\mu = 1\ (H^+,Na^+)ClO_4^-$	HS<0.01 in $CHCl_3$	66Ka
	$M^{+3} + 2.5\overline{(HS)_2} = \overline{MS_3(HS)_2} + 3H^+$	4.90	20	$\mu = 1\ (H^+,Na^+)ClO_4^-$	HS<0.01 in $CHCl_3$	66Ka
Eu^{+3}	$M^{+3} + 3S^- + 2\overline{HS} = \overline{MS_3(HS)_2}$	16.85	20	$\mu = 1\ (H^+,Na^+)ClO_4^-$	HS<0.01 in $CHCl_3$	66Ka
	$M^{+3} + 2.5\overline{(HS)_2} = \overline{MS_3(HS)_2} + 3H^+$	5.25	20	$\mu = 1\ (H^+,Na^+)ClO_4^-$	HS<0.01 in $CHCl_3$	66Ka
Tb^{+3}	$M^{+3} + 3S^- + 2\overline{HS} = \overline{MS_3(HS)_2}$	17.13	20	$\mu = 1\ (H^+,Na^+)ClO_4^-$	HS<0.01 in $CHCl_3$	66Ka
	$M^{+3} + 2.5\overline{(HS)_2} = \overline{MS_3(HS)_2} + 3H^+$	5.53	20	$\mu = 1\ (H^+,Na^+)ClO_4^-$	HS<0.01 in $CHCl_3$	66Ka
Tm^{+3}	$M^{+3} + 3S^- + 2\overline{HS} = \overline{MS_3(HS)_2}$	18.58	20	$\mu = 1\ (H^+,Na^+)ClO_4^-$	HS<0.01 in $CHCl_3$	66Ka
	$M^{+3} + 2.5\overline{(HS)_2} = \overline{MS_3(HS)_2} + 3H^+$	6.98	20	$\mu = 1\ (H^+,Na^+)ClO_4^-$	HS<0.01 in $CHCl_3$	66Ka
UO_2^{++}	$M^{++} + 2S^- + 2\overline{HS} = \overline{MS_2(HS)_2}$	14.32	20	$\mu = 1.0\ (H^+,Na^+)ClO_4^-$	HS<10^{-3}in $CHCl_3$	65K
	$M^{++} + 2\overline{(HS)_2} = \overline{MS_2(HS)_2} + 2H^+$	5.34	20	$\mu = 1.0\ (H^+,Na^+)ClO_4^-$	HS<10^{-3} in $CHCl_3$	65K

References

63K Krasovec, F. and Jan, J., Croat. Chem. Acta, 35, 183 (1963)

65K Krasovec, F. and Klofutar, C., J. Inorg. Nucl. Chem., 27, 2437 (1965)

66K Krasovec, F., Ostanek, M., and Klofutar, C., Anal. Chim. Acta, 36, 431 (1966)

66Ka Krasovec, F. and Klofutar, C., Radiochimica Acta, 6, 114 (1966)

67K Krasovec, F. and Klofutar, C., in "Solvent Extraction Chemistry", Eds. Dryssen, D., Liljenzin, J. O. and Rydberg, J., North-Holland, Amsterdam, p. 509 (1967)

68K Krasovec, F. and Klofutar, C., J. Inorg, Nucl. Chem., 30, 1640 (1968)

METAL ION, M	REACTION	LOG K	TEMP.	CONDITIONS, AQ.	CONDITIONS, ORG.	REF.
	Extractant Class: Phosphonates and phosphinates					
Extractant: Dibutyl butylphosphonate (DBB'P), S						
Reactions with acidic organic phosphate compounds						
	$\overline{S} + ((H)DE\text{-}HP)_2 = S\cdot((H)DE\text{-}HP)_2$	0.78	35	H_2SO_4 1.5	HDE-HP<0.4,S<0.5 in kerosene	63B/59B
		∿1.30	?		HDE-HP<0.01,S<1.0 in cyclohexane	63Za
	$\overline{(HS')_2} + \overline{S} = \overline{(HS')_2\cdot S}$	∿2.74			HS'<0.01,S<1.0 in cyclohexane	
	HS' = di para(1,1,3,3 tetramethylbutyl) phenyl phosphonic acid					
		∿1.30	?			
	HS' = 2-ethylhexyl, hydrogen 2-ethylhexylphosphonate					
	$\overline{S} + \frac{1}{2}\overline{((H)DE\text{-}HP)_2} = \overline{S((H)DE\text{-}HP)_2}$	∿0.17	35	H_2SO_4 1.5	S<0.4,HDE-HP<0.4 in kerosene	63B/59B
Reaction with diluents						
	$\overline{S} + \overline{CDCl_3} = \overline{S\cdot CDCl_3}$	0.2	?		S 0.04-1.5 in $CDCl_3$	64P
Ligand: NO_3^- , A^-						
H^+	$H^+ + A^- + \overline{S} = \overline{HA\cdot S}$	-2.66+1666/T		HA 1.0	S 0.05 in n-dodecane	59S

METAL ION, M	REACTION	LOG K	TEMP.	CONDITIONS, AQ.	CONDITIONS, ORG.	REF.
Ligand: NO_3^- , A^- (cont.)						
H^+	$H^+ + A^- + \overline{S} = \overline{HA \cdot S}$	ΔG 0.85	25	HA 1.00	S 0.05 in n-dodecane	59S
		ΔH -0.60				
		ΔS -5.3				
	$\overline{[HNO_3]}/a_W^{\,n}[HNO_3]$	0.55	25	?	S 0.2 in n-hexane	59Ba
	a_W = water activity	0.3	?	HA 0.5-7.3	S 0.7-1.1 in xylene	64S
		0.1	?	HA 0.1-1.4 , NaA 4-5	S 0.4-1.1 in xylene	
		-0.65	20	HA 0.1-6	S undil.	64Za
		-0.65	?	HA 0.12-13.8	S undil.	64Z
		0.0	?		S 10% in kerosene	
		0.0	?	HA→0	S→0 in CCl_4	63A
Zr^{+4}	$MO^{+2} + 2H^+ + 4A^- + 2\overline{S} = \overline{H_2MOA_4 \cdot 2S}$	2.37	25	μ = 3 $H^+(A^-, ClO_4^-)$	S<0.52 in CCl_4	62N
UO_2^{++}	$M^{+2} + 2A^- + 2\overline{S} = \overline{MA_2 \cdot 2S}$	ΔG -4.00	25	HA 1.00	S 0.55 in n-dodecane	59S
		ΔH 7.60				
		ΔS -39				
		1.98	25	μ = 0.714 HA	S 0.14 in CCl_4	59V
		2.81			S 0.19 in CCl_4	

METAL ION, M	REACTION	LOG K	TEMP.	CONDITIONS, AQ.	CONDITIONS, ORG.	REF.
Ligand: NO_3^- , A^- (cont.)						
UO_2^{++}	$M^{+2} + 2A^- + 2\overline{S} = \overline{MA_2 \cdot 2S}$	2.76	25	μ = 0.714 $H^+(A^-, ClO_4^-)$	S<0.14 in CCl_4	60P
		3.41	rt	MA 0.18	S<0.18 in CCl_4	63P
		4.71	?	HA → 0	S → 0 in CCl_4	63A
		3.89	22	HA 0.5	S 0.19 in benzene	63P/59H
		1.5	rt	MA_2 0.18	S<0.18 in chloroform	63P
		2.7	rt	MA_2 0.18	S<0.18 in dichloroethane	
		3.25	25	MA_2 0.2	S 0.5 in CCl_4	63P/58B
		3.46	25	μ = 0.714 HA	S<0.14 in CCl_4	63P/59V, 60P
		3.62	?	HA 0.5-2	S 0.5 in CCl_4	63P/60T, 60Pa
		4.85	?	HA → 0	S 0.5 in CCl_4	64R/58B
		5.32		HA 0.5	S 0.19 in benzene	
		2.27	20	HA 0.1-14	S undil.	64Za
		~4.0	?	MA 0.1, HA 1-2	S 0.3 in benzene	65N
NpO_2^{++}		0.60	20	HA 0.1-1	S undil.	64Za
Pu^{+4}	$M^{+4} + 4A^- + 2\overline{S} = \overline{MA_4 \cdot 2S}$	4.75	?	HA 0.5-2	S 0.5 in CCl_4	63P/60T, 60Pa

G

METAL ION, M	REACTION	LOG K	TEMP.	CONDITIONS, AQ.	CONDITIONS, ORG.	REF.
Ligand: NO_3^- , A^- (cont.)						
Pu^{+4}	$M^{+4} + 4A^- + 2\overline{S} = \overline{MA_4 \cdot 2S}$	4.61	25	HA>2.0	S 0.5 in CCl_4	63P/58B
		6.04	?	HA 2.0	S 0.5 in CCl_4	64R/61M
		4.22	20	HA 0.1-17	S undil.	64Za
PuO_2^{++}	$M^{+2} + 2A^- + 2\overline{S} = \overline{MA_2 \cdot 2S}$	2.37	25	HA 2.0	S 0.5 in CCl_4	63P/58B
		1.15	20	HA 0.1-15	S undil.	64Za
Am^{+3}	$M^{+3} + 3A^- + 3\overline{S} = \overline{MA_3 \cdot 3S}$	0.87	?	$\mu = 5\ (Na^+,H^+)ClO_4^-$	S<4 in kerosene	62Z
		0.87	?	HA 0.12-15.8	S undil.	64Z
		2.2	?		S 10% in kerosene	
Ligand: $SO_4^=$, $A^=$						
UO_2^{++}	$\overline{M(HS'_2)_2} + \overline{S} = \overline{M(HS'_2)_2 \cdot S}$ HS' = (H)DE-HP	2.78	32-34	H_2A 1.5	HS'<0.5 , S<0.5 in kerosene	63B/59B
Ligand: Cl^- , A^-						
La^{+3}	$M^{+3} + 3\overline{HS'} + \overline{S} = \overline{M(HS'_2)_3S}$ HS' = di para (1,1,3,3, tetramethylbutyl) phenyl phosphonic acid	~11	?	HA 0.3	S<1 , HS'<0.01 in cyclohexane	63Z
Eu^{+3}		~11.6	?			

METAL ION, M	REACTION	LOG K	TEMP.	CONDITIONS, AQ.	CONDITIONS, ORG.	REF.
Ligand: Cl^- , A^- (cont.)						
UO_2^{++}	$\overline{MA_2\cdot 2S} + \overline{S} = \overline{MA_2\cdot 3S}$	0.93	?	———	MA_2 0.02 , S 0.01-0.9 in CCl_4	63V
	$M^{+2} + 2\overline{(HS')_2} + \overline{S} = \overline{MS'_3{}^-HS'\cdot HS^+} + 2H^+$	∿7.36	?	HA 0.3	S<1.0 , HS'<0.01 in cyclohexane	63Za
	HS' = (H) DE-HP					
	HS' = di para (1,1,3,3 tetramethylbutyl) phenyl phosphonic acid	∿10	?	HA 0.3	HS'<0.01 , S<1.0	63Za
	HS' = 2-ethylhexyl, hydrogen 2-ethylhexyl phosphonate	∿6.74	?	HA 0.3	HS'<0.01 , S<1.0 in cyclohexane	
Am^{+3}	$M^{+3} + 3\overline{HS'} + \overline{S} = \overline{M(HS'_2)_3\cdot S}$	∿11	?	HA 0.3	S 1 , HS'<0.01 in cyclohexane	63Z
	HS' = di para(1,1,3,3 tetramethylbutyl) phenyl phosphonic acid					
Cm^{+3}		10.8	?	HA 0.3	S 1 , HS' 0.01 in cyclohexane	

References

58B Burger, L. L., J. Phys. Chem., 62, 590 (1958)

59B Blake, C. A., Horner, D. E. and Schmitt, J. M., USAEC Report ORNL 2259 (1959)

59Ba Baldwin, W. H., Higgins, C. E., and Soldano, B. A., J. Phys. Chem., 63, 118 (1959)

59H Healy, T. V. and Kennedy, J., J. Inorg. Nucl. Chem., 10, 128 (1959)

59S Siddall, T. H., J. Am. Chem. Soc., 81, 4176 (1959)

59V Voden, V. G., Nikitina, G. P. and Pushlenkov, M. P. Radiokhimiya, 1, 121 (1959)

60P Pushlenkov, M. P., Nikitina, G. P. and Voden, V. G., Radiokhimiya, 2, 21 (1960)

60Pa Petrov, K. A., Schevchenko, V. B., Timoshev, V. G., Maklyaev, F. A., Fokin, A. V., Redinov, A. V., Balandina, V. V., El'kina, A. V., Nagnibeda, Z. I. and Volkova, A. A., Zh. Neorg. Khim., 5, 498 (1960)

60T Timoshev, V. G., Petrov, K. A., Rodionov, A. A., Balandina, V. V., Volkova, A. A., El'kina, A. V. and Nagnibeda, Z. I., Radiokhimiya, 2, 419 (1960)

61M Martin, B., Ockenden, D. W. and Foreman, J. K., J. Inorg. Chem., 21, 96 (1961)

62N Nikitina, G. P. and Pushlenkov, M. F., Radiokhimiya, 4, 137 (1962)

62Z Zemlyanukhin, V. I., Savoskina, G. P. and M. F. Pushlenkov, Radiokhimiya, 4, 570 (1962)

63A Asano, M. and Nishi, T., J. Chem. Soc. Japan, Ind. Chem. Sec., 66, 1428 (1963)

63B Baes, C. F., Jr., Nucl. Sci. Eng., 16, 405 (1963)

63P Pushlenkov, M. F. and Shuvalov, O. N., Radiokhimiya, 5, 551 (1963)

63V Vdovenko, V. M., Lipovskii, A. A. and Nikitina, S. A., Radiokhimiya, 5, 585 (1963)

63Z Zangen, M., J. Inorg. Nucl. Chem., 25, 1051 (1963)

63Za Zangen, M., J. Inorg. Nucl. Chem., 25, 581 (1963)

64P Pushlenkov, M. F. and Komarov, E. V., Radiokhimiya, 6, 426 (1964)

64R Rozen, A. M. and Nikolotova, Z. I., Zh. Neorg. Khim., 9, 1725 (1964)

64S Sheppard, J. C., USAEC Report HW-81166 (1964)

64Z Zemlyanukhin, V. I., Savozhina, G. P. and Pushlenkov, M. F., Radiokhimiya, 6, 694 (1964)

64Za Zemlyanukhin, V. I., Savozhina, G. P. and Pushlenkov, M. F., Radiokhimiya, 6, 714 (1964)

65N Nikolaev, A. V., Gribanova, J. N., Yakovleva, N. I., Durasov, V. B., Khol'kina, I. D., Mironova, Z. N., Tsvetkov, E. N. and Kabachnik, M. I., Dokl. Akad. Nauk SSSR, 165, 578 (1965)

METAL ION, M	REACTION	LOG K	TEMP.	CONDITIONS, AQ.	CONDITIONS, ORG.	REF.
Extractant: Diisopentyl methyl phosphonate (DAM'P)						
Ligand: CNS^- , A^-						
H^+	$3(HA)_2 + 2\overline{S} = \overline{2(HA)_3 \cdot S}$	1.77	20 (?)	$NH_4A<1$, $H_2SO_4<0.5$	S 0.4 in benzene	60Y
Ligand: NO_3^- , A^- (cont.)						
H^+	$H^+ + A^- + \overline{S} = \overline{HA \cdot S}$	-0.52	19	HA<3.2	S 0.8 in benzene	59S
	$HA \cdot H_2O + \overline{S} + \overline{(HA \cdot H_2O)S}$	1.21	22	$\mu \sim 1$ and ~ 3	S undil.	60S
		1.34		$\mu \sim 1$ and ~ 3	S 2-30% in benzene	
	$HA \cdot 3H_2O + \overline{S} = \overline{(HA \cdot 3H\ O)S}$	1.03		$\mu \sim 1$ and ~ 3	S 2-30% in CCl_4	
		1.21		$\mu \sim 1$ and ~ 3		
	$2HA \cdot 0.5H_2O + \overline{S} = \overline{(HA \cdot 0.5H_2O)_2S}$	0.0		$\mu \sim 1$ and ~ 3	S undil.	
		-0.4		$\mu \sim 1$ and ~ 3	S 2-30% in CCl_4	
		0.14		$\mu \sim 1$ and ~ 3	S 2-30% in benzene	
		0.0		$\mu \sim 1$ and ~ 3	S 2-30% in kerosene	
	$H^+ + A^- + \overline{S} = \overline{HAS}$	-0.40	25	HA 0.3-3	S 0.5 in CCl_4	64R
		0.34	?	HA 0.2-9.4	S 50% in m-xylene	64Z
		0.26	?	HA 0-10	S 50% in kerosene	
		1.24	?	HA 0.2-13	S 50% in dibutylether	

METAL ION, M	REACTION	LOG K	TEMP.	CONDITIONS, AQ.	CONDITIONS, ORG.	REF.
Ligand: NO_3^- , A^- (cont.)						
H^+	$H^+ + A^- + \overline{S} = \overline{HAS}$	0.30	?	HA 0.2-12	S 50% in benzene	64Z
		0.26	?	HA 0.3-9.7	S 50% in CCl_4	
		-0.48	?		S 50% in $CHCl_3$	
		0.5	?	HA 0.10	S undil.	64Z/62Z
		-0.04			S 30% in kerosene	64Z
		-0.23	?		S 20% in kerosene	
		-0.26	?		S 10% in kerosene	
		-0.27	25	HA 2	S<1.5 in benzene	65S
Ce^{+3}	$M^{+3} + 3A^- + 2\overline{S} = \overline{MA_3 \cdot 2S}$	1.8	20	$\mu = 2$ $(Li^+,H^+)A^-$, $LiClO_4$	S<0.34 in kerosene	62U
		0.9 to 1.6	20		S<0.34 in benzene or xylene	
		1.55	20	$\mu \rightarrow 0$	S<0.34 in benzene	
	$M^{+3} + 3A^- + 3\overline{S} = \overline{MA_3 \cdot 3S}$	0.96	rt	HA 0.01 , CaA_2 1.26	S 0.4 in toluene	65A
		0.85	rt	HA 0.01 CaA_2 3.5	S 0.4 in toluene	
		1.82	25	HA 2, $\mu = 2$	S<0.3 in benzene	65O
		1.82	20	AlA_3 2.0 , HA 2.0	S in benzene	65Sb

METAL ION, M	REACTION	LOG K	TEMP.	CONDITIONS, AQ.	CONDITIONS, ORG.	REF.
Ligand: NO_3^- , A^- (cont.)						
Zr^{+4}	$MO^{+2} + 2H^+ + 4A^- + 2\overline{S} = \overline{H_2MOA_4 \cdot 2S}$	3.50	20	HA 0.5 , μ = 2 LiA	S 0.65 in kerosene	65U
	$MO^{+2} + 2H^+ + 4A^- + \overline{S} = \overline{H_2MOA_4 \cdot S}$	2.8	20		S 0.8 in kerosene	
Th^{+4}	$M^{+4} + 4A^- + 2\overline{S} = \overline{MA_4 \cdot 2S}$	2.92	25	HA<3	S 0.4 in benzene	64A
		3.77	22	HA 0.5 μ = 2, Li(NO_3^-,ClO_4)	S 0.3-0.01 in p-xylene	65Ua
	$M^{+4} + 4A^- + 3\overline{S} = \overline{MA_4 \cdot 3S}$	4.49	22			
UO_2^{++}	$M^{+2} + 2A^- + \overline{S} = \overline{MA_2S}$	2.87	?	HA 1, MA_2 0.07	S<0.075 in hydrogenated kerosene	62F
	$M^{+2} + 2A^- + 2\overline{S} = \overline{MA_2 \cdot 2S}$	4.40	19	HA 0.05	S 0.8 in kerosene	59S
		2.42	?	μ = 1 NH_4A 1, HA 0.05	S<0.1 in hydrogenated kerosene	62F
		~3.11	25	HA 0.05	S undil.	64R
		3.25	25		S 0.5 in CCl_4	
		3.4-3.5	25		S 0.5 in hexane	
		3.7-3.8	25		S 0.5 in benzene	
		2.84-3.18	?	HA 2	S<0.1 in benzene	62S
		3.24	25	MA_2 0.01-0.2 + HA	S 0.5 in CCl_4	65R
		3.02	40			
		2.78	60			

METAL ION, M	REACTION	LOG K	TEMP.	CONDITIONS, AQ.	CONDITIONS, ORG.	REF.
Ligand: NO_3^- , A^- (cont.)						
UO_2^{++}	$M^{+2} + 2A^- + \overline{S} = \overline{TBP} = \overline{MA_2 \cdot S \cdot TBP}$	2.04	?	μ = 1 NH_4A 1, HA 0.05	S<0.1 TBP<0.8 in hydrogenated kerosene	62F
		2.33	?	HA 1, MA_2 0.07	S<0.075 TBP 0.74 in hydrogenated kerosene	
	$M(DiPP)_2 + \overline{S} = \overline{M(DiPP)_2 \cdot S}$	4.0	?	μ = 2 HA	S<0.1 in benzene	62S
	$MS'_2 + \overline{S} = \overline{MS'_2 \cdot S}$	3.72	?	μ = 2 HA	S<0.1 in benzene	
	HS' = Isopenthyl methyl phosphoric acid					
Am^{+3}	$M^{+3} + 3A^- + 3\overline{S} = \overline{MA_3 \cdot 3S}$	0.92	20	NaA 5.0	S 0.03-0.8 in kerosene	62Z
		0.92	?	HA 0-10	S undil.	64Z
		1.49	?		S 30% in kerosene	
		1.74	?		S 20% in kerosene	
		1.97	?		S 10% in kerosene	
		1.0	?	HA 0.2-9.4	S 50% in m-xylene	
		1.06	?	HA 0-10	S 50% in kerosene	
		0.98	?	HA 0.2-13	S 50% in dibutylether	
		0.71	?	HA 0.2-12	S 50% in benzene	
		0.5	?	HA 0.3-9.7	S 50% in CCl_4	
		-1	?		S 50% in $CHCl_3$	

METAL ION, M	REACTION	LOG K	TEMP.	CONDITIONS, AQ.	CONDITIONS, ORG.	REF.
Ligand: Cl^- , A^-						
H^+	$H^+ + A^- + \overline{S} = \overline{HAS}$	1.26	?	HA 0.2	S 0.5-0.1 in m-xylene	65Sa
Hf^{+4}	$M^{+4} + 2H_2O + 2A^- + 2\overline{S} + 2H^+ = \overline{M(OH)_2A_2 \cdot 2S}$	0.31	?	A^- 0.05	S 0.04-0.3 in m-xylene	
U^{+4}	$M^{+4} + 4A^- + 2\overline{S} = \overline{MA_4 \cdot 2S}$	$K_{DAM'P}/K_{TBP}$ = 2.48	?	HA<6.0	S<0.36 in CCl_4	61S
	$\overline{MA_4S_2} + \overline{S} = \overline{MA_4S_3}$	2.07	?	———	S 0-0.127 , MA_4 0.05 in CCl_4	65L
UO_2^{++}	$M^{+2} + 2A^- + 2\overline{S} = \overline{MA_2 \cdot 2S}$	$K_{DAM'P}/K_{TBP}$ = 2.07	?	HA 5.0	S 0.77 in CCl_4	61S
	$\overline{MA_2 \cdot 2S} + \overline{S} = \overline{MA_2 \cdot 3S}$	0.94	?	———	S<0.8 in CCl_4	63V
Ligand: I^- , A^-						
Li^+	$M^+ + A^- + 8H_2O = \overline{(M^+ + A^-) \cdot 8H_2O}$	-1.02	?	MA 0.1 - sat.	S undil.	67R
	$M^+ + A^- + 2.4\ H_2O + 3\overline{S} = \overline{MA \cdot 2.4\ H\ O \cdot 3S}$	-2.42	?		S 0.5 in benzene	
	$M^+ + A^- + 2.4\ H_2O = \overline{MA \cdot 2.4\ H_2O}$	-3.33	?		S 0.5 in benzene	
Na^+	$M^+ + A^- + 10\ H_2O = \overline{(M^+ + A^-)\ 10\ H_2O}$	-1.52	?		S undil.	
	$M^+ + A^- + 3H_2O = \overline{MA \cdot 3H_2O}$	3.4	?		S 0.5 in benzene	
	$M^+ + A^- + 3H_2O + 2\overline{S} = \overline{MA \cdot 3H_2O \cdot 2S}$	-3.8	?		S 0.5 in benzene	
K^+	$M^+ + A^- + 10\ H_2O = \overline{(M^+ + A^-) \cdot 10\ H_2O}$	- 1.57	?		S undil.	
Cs^+	$M^+ + A^- + 9H_2O = \overline{(M^+ + A^-) \cdot 9H_2O}$	-1.48	?		S undil.	

References

59S Shevchenko, V. B., Slepchenko, I. G., Shmidt, V. S. and Nenarokomov, E. A., Atomnaya Energiya, 7, 236 (1959)

60S Solovkin, A. S., Zh. Neorg. Khim., 5, 1345 (1960)

60Y Yagodin, G. A. and Mostovaya, O. A., Zh. Priklad. Khim., 33, 2459 (1960)

61S Shevchenko, V. B., Schmidt, V. S. and Nenarokomov, E. A., Radiokhimiya, 3, 129 (1961)

62F Fomin, V. V., Karturkova, R. E. and Maiorova, E. P., Ekstraktsiya Teoriya, Primenenie, Apparatura, Sb. Statei, No. 2, 37-46 (1962)

62S Shevchenko, V. B., Smelov, V. S. and Strakhova, A. V., Zh. Neorg. Khim., 7, 1736 (1962)

62U Ukraintsev, E. V. and Kashchew, N. F., Radiokhimiya, 4, 279 (1962)

62Z Zemlyanukhin, V. I., Savoskina, G. P. and Pushlenkov, M. F., Radiokhimiya, 4, 655 (1962)

63V Vdovenko, V. M. and Lipovskii, A. A., Radiokhimiya, 5, 585 (1963)

64A Alian, A., Proc. Int. Conf. Peaceful Uses Atom. Energy, Genega, 1964, United Nations, 12, 256 (1965)

64R Rozen, A. M. and Nikolotova, Z. I., Zh. Neorg. Khim., 9, 1725 (1964)

64Z Zemlyanukhin, V.I., Sarovskina, G. P. and Pushlenkov, M. F., Radiokhimiya, 6, 694 (1964)

65Q Alian, A. and Moustapha, Z., Indian J. Chem., 3, 390 (1965)

65L Lipovskii, A. A., Nikitina, S. A. and Yakovlev, N. E., Radiokhimiya, 7, 563 (1965)

65O Orlov, Yu. F., Iovin, B. I. and Shvedov, V. P., Zh. Obshchei Khim., 35, 2046 (1965)

65R Rozen, A. M., Nikolotova, Z. I., Petrov, K. A., Skotnikov, A. S. and Teterin, E. G., Radiokhimiya, 7, 517 (1965)

METAL ION, M	REACTION	LOG K	TEMP.	CONDITIONS, AQ.	CONDITIONS, ORG.	REF.

65S Shvedov, V. P. and Orlov, Yu. F., Zh. Neorg. Khim., 10, 2774 (1965)

65Sa Sinegribova, O. A. and Yagodin, G. A., Zh. Neorg. Khim., 10, 1250 (1965)

65Sb Shvedov, V. P. and Orlov, Yu. F., Zh. Prikl. Khim., 38, 1605 (1965)

65U Ukraintsev, E. V., Radiokhimiya, 7, 648 (1965)

65Ua Ukraintsev, E. V., Radiokhimiya, 7, 641 (1965)

67R Rozen, A. M. and Mikhailichenko, A. I., Zh. Neorg. Khim., 12, 741 (1967)

Extractant: Octyl Dioctyl Phosphinate,(ODO'P), S

Ligand: NO_3^- , A^-

METAL ION, M	REACTION	LOG K	TEMP.	CONDITIONS, AQ.	CONDITIONS, ORG.	REF.
H^+	$H^+ + A^- + \overline{S} = \overline{HA \cdot S}$	0.60	25	HA 0.05	S 0.5 in CCl_4	64R
UO_2^{++}	$M^{+2} + 2A^- + 2\overline{S} = \overline{MA_2 2S}$	4.92	25	HA 0.05		
		3.7-5.9	25	MA_2 0.02-0.2 + HA	S 0.5 in CCl_4	65R
		4.7-5.7	40			
		4.3-5.3	60			

64R Rozen, A. M. and Nikolotova, Z. I., Zh. Neorg. Khim., 9, 725 (1964)

65R Rozen, A. M., Nikolotova, Z. I., Petrov, K. A., Skotnikov, A. S. and Teterin, E. G., Radiokhimiya, 7, 517 (1965)

METAL ION, M	REACTION	LOG K	TEMP.	CONDITIONS, AQ.	CONDITIONS, ORG.	REF.
		Extractant Class: Phosphine oxides				
Extractant: Tributyl phosphine oxide (TB'PO), S						
Distribution constants						
	$S = \overline{S}$	0.73-1.35	25	H_2O	S 0.015m to 1.5m in CCl_4	60H
		0.23-0.09	27-50	H_2O	S 100% (from solubility)	60Ha
		0.58-0.28	21	HCl 0.3-2.07	S<2% in toluene	61I
		1.06	25	H_2O	S 0.1 in benzene	65C,66G
		-0.3	25		S 0.1 in hexane	
		0.78	25		S 0.1 in toluene	
		-0.22	25		S 0.1 in cyclohexane	
Interaction with acidic org. phosph. compounds						
	$\overline{S} + \frac{1}{2}\overline{H_2(DE\text{-}HPA)_2} = \overline{S\cdot(HDE\text{-}HPA)}$	0.92	25	H_2SO_4 1.5	S<0.4 HDE-HPA 0.4 in kerosene	65B/59B
	$\overline{S} + \overline{H_2(DE\text{-}HPA)_2} = \overline{S\cdot H_2(DE\text{-}HPA)_2}$	1.52	25		HDE-HPA in kerosene	
	$\overline{S} + \overline{(H)DpCl\text{-}PhP} = \overline{S\cdot(H)DpCl\text{-}PhP}$	4.98	?	$HClO_4$ 0.1	(H)DpCl-PhP<<S < 10^{-3} in CCl_4	66K,67K
Reaction with diluents						
	$\overline{S} + \overline{CDCl_3} = \overline{S\cdot CDCl_3}$	0.68	?		S 0.04-0.66 in $CDCl_3$	64P

METAL ION, M	REACTION	LOG K	TEMP.	CONDITIONS, AQ.	CONDITIONS, ORG.	REF.
Ligand: NO_3^- , A^-						
H^+	$H^+ + A^- + \overline{S} = \overline{HAS}$	1.23	25	HA<1	S<0.2 in benzene	64D,66G
		1.30			S<0.2 in toluene	
		1.04		HA<1.5	S<0.2 in n-hexane	
	$H^+ + A^- + H_2O + \overline{HAS} = \overline{(2HA)\cdot SH_2O}$	-2.55		HA 3-7	S<0.2 in benzene	
	$H^+ + A^- + S(H_2O)_x = S\cdot HA + xH_2O$	1.3		HA<8	S 0.1 in toluene	65C
		1.04			S 0.1 in hexane	
	$H^+ + A^- + \overline{S} = \overline{HA\cdot S}$	0.73	?	HA → 0	S → 0 in CCl_4	63A
	$S(H_2O)_x + \overline{HA} = S.HA + xH_2O$	5.64	?	HA<6	S 0.4 in benzene	
UO_2^{++}	$M^{+2} + 2A^- + 2\overline{S} = \overline{MA_2.2S}$	6.58	25	HA<0.216	S<1.8 10^{-2} in CCl_4	59V
		6.53		(HA + $HClO_4$) 0.71	S 8.4 10^{-3} in CCl_4	60P
		6.4	25	μ→0,(1-13 HA)	S 0.73 in CCl_4	61N
		4.63-4.98	?	HA 2	S<0.01 in benzene	62S
		5.71	?	HA → 0	S→0 in CCl_4	63A
		6.7-7.3	rt	MA_2 0.18	S 10^{-2}-10^{-4} in CCl_4	63P
		5.92	22	HA 0.5	S 0.19 in benzene	63P/59H

METAL ION, M	REACTION	LOG K	TEMP.	CONDITIONS, AQ.	CONDITIONS, ORG.	REF.
Ligand: NO_3^- , A^- (cont.)						
UO_2^{++}	$M^{+2} + 2A^- + 2\overline{S} = \overline{MA\ .2S}$	6.50	?	HA 0.5-2	S 0.5 in CCl_4	63P/60T, 60Pa
		6.6	rt	MA_2 0.18	S 10^{-2}-10^4 in dichloroethane	63P
		4.8			S 10^{-2}-10^4 in $CHCl_3$	
		7.23	25	HA 0.72 $\mu = 0.71$	S 8.4 10^{-3} in CCl_4	63P/60P, 59V
		7.13	25	MA_2 0.2	S 0.5 in CCl_4	63P/58B
		7.2	25	HA $\rightarrow$ 0	S 0.5 in CCl_4	64R/58B
		6.61	25	HA 0.5	S 0.19 in benzene	64R/59H
		~8.5	?	MA_2 0.1 HA 1-2	S 0.3 in benzene	65N
		5.28	?	HA 1.0	$S<10^{-2}$ in CCl_4	67K
	$M^{+2} + A^- + ClO_4^- + 2\overline{S} = \overline{MAClO_4.2S}$	5.67	25	$(HA + HClO_4)$ 0.71	S 8.4 x 10^{-3} in CCl_4	60P
	$\overline{M(DiPP)_2} + \overline{S} = \overline{M(DiPP)_2}$	5.1	?	$\mu = 2$ (HA)	S<0.01 in benzene	62S
	$\overline{MS'_2} + \overline{S} = \overline{MS'_2.S}$	5.08	?			
	HS' = Isopentyl methyl phosphoric acid					
Pu^{+4}	$M^{+4} + 4A^- + 2\overline{S} = \overline{MA_4.2S}$	7.40	?	HA 0.5-2	S 0.5 in CCl_4	63P/60T, 60Pa

METAL ION, M	REACTION	LOG K	TEMP.	CONDITIONS, AQ.	CONDITIONS, ORG.	REF.
Ligand: NO_3^- , A^- (cont.)						
Pu^{+4}	$M^{+4} + 4A^- + 2\overline{S} = \overline{MA_4.2S}$	7.85	?	HA 2.0	S 0.5 in CCl_4	64R/61M
Am^{+3}	$M^{+3} + 3A^- + 3\overline{S} = \overline{MA_3.3S}$	3.25	?	μ = 5 $(Na^+,H^+)A^-$, $NaClO_4$	S<0.4 in kerosene	62Z
Ligand: H_2O						
	$H_2O + \overline{S} = \overline{S.H_2O}$	0.74	?	H_2O	S<0.07 in CCl_4	66R
		0.58	?		S 0.07 in CS_2	
	$2\overline{S.H_2O} = \overline{(S.H_2O)}$	-0.36	?		0.1<S<0.2 in CCl_4	
Ligand: $SO_4^=$, $A^=$						
UO_2^{++}	$\overline{M(HS'_2)_2} + \overline{S} = \overline{M(HS'_2)_2.S}$ S' = (H)DE-HP	4.28	25	H_2A 1.5	HS'<0.5, S<0.5 in kerosene	63B/59B
Ligand: Cl^- , A^-						
U^{+4}	$\overline{MA_4S_2} + \overline{S} = \overline{MA_4.S_3}$	1.93	?	———	M^{+4} 0.05; S 0-0.4 in kerosene	65L
TcO_4^-	$H^+ + TcO_4^- + \overline{S} = \overline{HTcO_4.S}$	2.44	?	HA 1.0	S 0.1-0.003 in CH_2Cl_2	67K
ReO_4^-	$H^+ + ReO_4^- + \overline{S} = \overline{HReO_4.S}$	2.16	?	HA 1.0	S 0.1-0.003 in CH_2Cl_2	67K
Ligand: ClO_4^- , A^-						
UO_2^{++}	$M^{+2} + 2A^- + 2\overline{S} = \overline{MA_2.2S}$	5.30	25	HA 0.71	S $8.4 \cdot 10^{-3}$ in CCl_4	60P

References

58B Burger, L.L., J. Phys. Chem., 62, 590 (1958)

59B Blake, C. A., Horner, D. E. and Schmitt, J. N., USAEC Report ORNL 2259 (1959)

59H Healy, T. V. and Kennedy, J., J. Inorg. Nucl. Chem., 10, 128 (1959)

59V Voden, V. G., Nikitina, G. P. and Pushlenkov, M. P., Radiokhimiya, 1, 121 (1959)

60H Higgins, C. E. and Baldwin, W. H., Anal. Chem., 32, 233 (1960)

60Ha Higgins, C. E., and Baldwin, W. H., Anal. Chem., 32, 236 (1960)

60P Pushlenkov, M. P., Nikitina, G. P. and Voden, V. G., Radiokhimiya, 2, 215 (1960)

60Pa Petrov, K. A. et al., Zh. Neorg. Khim., 5, 498 (1960)

60T Timoshev, V. G. et al., Radiokhimiya, 2, 419 (1960)

61I Ishimori, T. and Fujino, T., Nippon Genshiryoku Gakkaishi, 3, 276 (1961)

61M Martin, B., Ockenden, D. W. and Foreman, J. K., J. Inorg. Nucl. Chem., 21, 96 (1961)

61N Nomura, S. and Hara, R., Anal. Chim. Acta, 25, 212 (1961)

62S Shevchenko, V. B., Smelov, V. S. and Strakhova, A. V., Zh. Neorg. Khim., 7, 1736 (1962)

62Z Zemlyanukhin, V. I., Saroskina, G. P. and Pushlenkov, M. F., Radiokhimiya, 4, 570 (1962)

63A Arano, M., and Nishi, T., J. Chem. Soc. Japan, Ind. Chem. Sec., 66, 1428 (1963)

63B Baes, C. F., Jr., Nucl. Sci. Eng., 16, 405 (1963)

63P Pushlenkov,M. F. and Shuvalov, O. N., Radiokhimiya, 5, 551 (1963)

64D Duyckaerts, G. and Goffart, J., EURATOM Report EUR-437 (1964)

64P Pushlenkov, M. F. and Komarov, E. V., Radiokhimiya, 6, 426 (1964)

64R Rozen, A. M. and Nikolotova, Z. I., Zh. Neorg. Khim., 9, 1725 (1964)

65C Caligara, F., Dreze, Ph., Duyckaerts, G., Fuger, J., Goffart, J. and Müller, W., EURATOM Report EUR-2297 e. (1965)

65L Lipovskii. A. A., Nikitina, S. A. and Yakovlev, N. E., Radiokhimiya, 7, 563 (1965)

65N Nikolaev, A. V., Gribanova, I. N., Yakovleva, N. I., Durasov, V. B., Khol'kina, I. D., Mironova, Z. M., Tsvatkov, E. N. and Kabachnik, M. I., Dokl. Akad. Nauk SSSR, 165, 578 (1965)

66G Goffart, J. and Duyckaerts, G., Anal. Chim. Acta, 36, 499 (1966)

66R Roland, G. and Duyckaerts, G., Spectrochim. Acta, 22, 793 (1966)

67K Krasovec, F. and Klofutar, C., in "Solvent Extraction Chemistry", Eds. Dyrssen, D., Liljenzin, J. O. and Rydberg, J. North-Holland, Amsterdam, p. 509 (1967)

68K Krasovec, F. and Klofutar, C., J. Inorg. Nucl. Chem., 30, 1640 (1968)

METAL ION, M	REACTION	LOG K	TEMP.	CONDITIONS, AQ.	CONDITIONS, ORG.	REF.
Extractant: Trioctyl phosphine oxide (TO'PO), S						
Distribution constants						
	$S = \overline{S}$	>5	25	H_2O	S 0.1 in benzene	65Ca
Reaction with acidic organic phosphorus compounds						
	$\overline{S} + \overline{(HS')_2} = \overline{S\cdot(HS')_2}$	1.7 to 2.3	?	——	TO'PO + HS' in kerosene	63I/58B
	HS' = (H)DE-HP	1.7 to 2.2	22	——	$(HS')_2$ + TO'PO	62B
		1.4	22	——	$(HS')_2$ + TO'PO 0.1 and 0.2 in octane	62B
	$\overline{HDBP} + \overline{S} = \overline{HDBP\cdot S}$	4.82	25	H_2SO_4 0.1	HDBP<0.1 + S<0.1 in hexane	66L,67L, 68L
		4.29			HDBP<0.1 + S<0.1 in CCl_4	
	$\overline{S} + \overline{(HS')_2} = \overline{S(HS')_2}$	∿2.11	?	HCl 0.3	HS' 0.1 + S<1.0 in cyclohexane	63Z
	HS' = 2-ethylhexyl hydrogen-2-ethylhexylphosphonate					
	$2S + \overline{(HS')_2} = \overline{2S\cdot HS'}$	2.9	?		S 0.1-0.2 + HS' 2.10^{-3} in heptane	66W
	HS' = Dinonyl naphthalene sulphonic acid					
	$\overline{S} + \overline{(H)DpCl\text{-}PhP} = \overline{S\cdot(H)DpCl\text{-}PhP}$	4.98	rt	$HClO_4$ 0.1	(H)DpCl-PhP <3.5 x 10^{-} <<S in CCl_4	68K

METAL ION, M	REACTION	LOG K	TEMP.	CONDITIONS, AQ.	CONDITIONS, ORG.	REF.
Ligand: NO_3^- , A^-						
H^+	$H^+ + A^- + \overline{S^-} = \overline{HA.S}$	0.86	12.4	μ = 2 $(H^+,Na^+)A^-$ H^+<0.4	S<0.05 in cyclohexane	60Z
		0.96	25			
		1.06	35			
		1.15	10.1	HA 7.5 . 10^{-3}, NaA 0.6	S 0.04 in cyclohexane	63S
		0.97	25			
		0.69	55.0			
		ΔH -3.9	10-55			
		ΔS -8.8				
		1.22	25	HA<1	S<0.1 in benzene	64D
		1.18	25	HA 0.01 , NaA 2	S<0.1 in benzene	
		0.99	25	HA 8.10^{-2} , NaA 0.6	S 0.04 in cyclohexane	63S,64H
		ΔH -3.9	10-55			
		ΔS 8.8				
		1.0	25	HA 1	S<0.1 in 1,2 dichlorobenzene	64M
		1.14	25	HA 0.05	S 0.05 in CCl_4	64R
		1.18		HA 0.1-1.26	S 0.1 in benzene	67G
		1.35		HA<0.01 , NaA 2, μ = 2	S<0.05 in n-octane	67G

METAL ION, M	REACTION	LOG K	TEMP.	CONDITIONS, AQ.	CONDITIONS, ORG.	REF.
Ligand: NO_3^- , A^- (cont.)						
Ce^{+3}	$M^{+3} + 3A^- + 3\overline{S} = \overline{MA_3.3S}$	3.7	rt	HA 0.2 , LiA<1.2	S 0.4 in toluene	65A,64A
Th^{+4}	$M^{+4} + 4A^- + 3\overline{S} = \overline{MA_4.3S}$	7.7-8.3	12	μ = 2 (Na^+,H^+) A^-, H^+<0.01	S<0.1 in cyclohexane	60Z
		7.78	25			
		8.0	35			
		10.27	25	KA 0.1, HA 10^{-3}	S<0.08 in cyclohexane	64P,65P
		9.87	25	HA 0.1	S<0.05 in 1.2 dichlorobenzene	
UO_2^{++}	$M^{+2} + 2A^- + 2\overline{S} = \overline{MA_2.S_2}$	8.62	9	HA 0.022	S 0.025 in cyclohexane	63S
		8.42	25			
		8.07	49.95			
		ΔH -5.6	25			
		ΔS 19.7	25			
		8.43	25	HA 0.023	S 0.025 in cyclohexane	64H
		ΔH -5.8	10-55			
		ΔS 19	10-55			
		7.11	25	HA 0.05	S 0.5 in CCl_4	64R

METAL ION, M	REACTION	LOG K	TEMP.	CONDITIONS, AQ.	CONDITIONS, ORG.	REF.
Ligand: NO_3^- , A^- (cont.)						
UO_2^{++}	$M^{+2} + 2A^- + 2\overline{S} = \overline{MA_2.S_2}$	7.4	10	$MA_2<0.1$, HA 0.05	S 0.5 in CCl_4	65R
		7.1	25			
		6.4	50			
Pu^{+4}	$M(IV) + 2\overline{S} = \overline{M(IV)S_2}$	5.10	~20	HA 6.0	S<0.1 in cyclohexane	61M
PuO_2^{++}	$M(VI) + 2\overline{S} = \overline{M(VI)S_2}$	4.50	~20	HA 0.69	S<0.1 in cyclohexane	
Am^{+3}	$M^{+3} + 3A^- + 3\overline{S} = \overline{MA_3.3S}$	4.78	~20	HA 0.01 , NH_4A 1.0	S 5% in toluene	62I
	$M^{+3} + 3A^- + \overline{S(H_2O)_x} = \overline{MA_3.3S} + xH_2O$	6.34	25	HA 0.01 , NaA 2	S<0.1 in n-octane	65Ca
Cm^{+3}		6.11	25			
Tc(VII)	$H^+ + MO_4^- + 2\overline{S} = \overline{HMO_4.2S}$	1.04	25	HA 2	S<0.3 in hydrocarbon	60B
Ligand: H_2O						
	$H_2O + \overline{S} = \overline{H_2O.S}$	-0.25	23	$HClO_4<0.05$	S<1 in CCl_4	65C
Ligand: $SO_4^=$, $A^=$						
UO_2^{++}	$\overline{M(DE\text{-}HP)_2}\ \overline{((H)DE\text{-}HP)_2} + \overline{S} = \overline{M(DE\text{-}HP)_2.((H)DE\text{-}HP)_2.S}$	<3.84	?	HA 0.5	S<0.01 , (H)DE-HP 0.1 in kerosene	63I
	$\overline{M(DE\text{-}HP)_2((H)DE\text{-}HP)\ (H_2O)_m} + \overline{S} = \overline{M(DE\text{-}HP)_2((H)DE\text{-}HP)_2(H_2O)_{m-1}.S} + H_2O$	<1.74	?			

METAL ION, M	REACTION	LOG K	TEMP.	CONDITIONS, AQ.	CONDITIONS, ORG.	REF.
Ligand: NO_3^- , A^- (cont.)						
Ce^{+3}	$M^{+3} + 3A^- + 3\overline{S} = \overline{MA_3.3S}$	3.7	rt	HA 0.2 , LiA<1.2	S 0.4 in toluene	65A,64A
Th^{+4}	$M^{+4} + 4A^- + 3\overline{S} = \overline{MA_4.3S}$	7.7-8.3	12	μ = 2 ($Na^=$,H^+) A^- , H^+ 0.01	S<0.1 in cyclohexane	60Z
		7.78	25			
		8.0	35			
		10.27	25	KA 0.1 , HA 10^{-3}	S<0.08 in cyclohexane	64P,65P
		9.87	25	HA 0.1	S<0.05 in 1.2 dichloro-benzene	
UO_2^{++}	$M^{+2} + 2A^- + 2\overline{S} = \overline{MA_2.S_2}$	8.62	9	HA 0.022	S 0.025 in cyclohexane	63S
		8.42	25			
		8.07	50			
		ΔH-5.6	9-55			
		ΔS19.7				
		8.43	25	HA 0.023	S 0.025 in cyclohexane	64H
		ΔH-5.8	10-55	HA 0.023	S 0.025 in cyclohexane	64H
		ΔS19	10-55			
		7.11	25	HA 0.05	S 0.5 in CCl_4	64R

METAL ION, M	REACTION	LOG K	TEMP.	CONDITIONS, AQ.	CONDITIONS, ORG.	REF.
Ligand: $SO_4^{=}$, $A^{=}$ (cont.)						
UO_2^{++}						
	$\overline{M(DE\text{-}HP)_2((H)DE\text{-}HP)_2} + \overline{S} = \overline{M(DE\text{-}HP)_2} \cdot \overline{((H)DE\text{-}HP).S} + \frac{1}{2}\ \overline{((H)DE\text{-}HP)_2}$	3.12	?	HA 0.5	S<0.01, (H)DEHP 0.1 in kerosene	63I
	$M^{+2} + 2\overline{(H)DBP} + \overline{S} = \overline{M(DBP)_2.S} + 2H^{+}$	8.18	25	H_2A 0.1	$(H)DBP \geq 10^{-4}$, S<0.05 in hexane	66L
		8.66	25	H_2A 0.1	$[(H)DBP]^2[S]^- > 10^{-7.4}$, S<0.06 in CCl_4	
	$M^{+2} + 3\overline{(H)DBP} + \overline{S} = \overline{M(DBP)_2((H)DBP \cdot S} + 2H^{+}$	11.95	25	H_2A 0.1	$(H)DBP\ 10^{-2.8}$, $S<4.7.10^{-3}$ in CCl_4	
Ligand: F^- , A^-						
Hf^{+4}	$MA_2^{+2} + 2ClO_4^- + \overline{S} = \overline{MA_2(ClO_4)_2S}$	10.05	25	$HClO_4$ 3 , HA<0.8	S<0.05 in cyclohexane	64V
	$MA_2^{+2} + 2ClO_4^- + 2\overline{S} = \overline{MA_2(ClO_4)_2S_2}$	12.01	25			
	$M^{+4} + 2A^- + 2ClO_4^- + 1.5\overline{S} = \overline{MA_2(ClO_4)_2} \cdot \overline{1.5\ S}$	11.39	25			
Ta^{+5}	$MA_4^{+} + ClO_4^- + 2\overline{S} = \overline{MA_4(ClO_4).2S}$	0.0	25	$A^- < 10^{-5}$, $HClO_4$ 3	S 0.01 in $CHCl_3$	65V
Ligand: Cl^- , A^-						
UO_2^{++}	$M^{+2} + 2\overline{(HS')_2} + S^- = \overline{MS'_3{}^- \cdot HS'HS^+} + 2H^{+}$	8.74	?	HA 0.3	HS'<0.01 S<1.0 in cyclohexane	63Z
	HS' = 2-ethylhexyl hydrogen 2-ethylhexylphosphonate					

METAL ION, M	REACTION	LOG K	TEMP.	CONDITIONS, AQ.	CONDITIONS, ORG.	REF.
Ligand: Cl^- , A^- (cont.)						
Tc(VII)	$H^+ + MO_4^- + 2\overline{S} = \overline{HMO_4 \cdot 2S}$	2.53	25	HA 1	S<0.2 in CCl_4	60B
Ligand: ClO_4^- , A^-						
Hf^{+4}	$M^{+4} + 4\ ClO_4^- + \overline{S} = \overline{M(ClO_4)_4S}$	2.77	25	$HClO_4$ 3 , HA<0.8	S<0.05 in cyclohexane	64V
	$M^{+4} + 4ClO_4^- + 2\overline{S} = \overline{M(ClO_4)_4S_2}$	4.32	25			
	$M^{+4} + 4ClO_4^- + 1.5\ \overline{S} = \overline{M(ClO_4) \cdot 1 \cdot 5S}$	3.93	25			
Zn^{+2}	$M^{+2} + 0.25\overline{(NaD)_8} + 1.65\ \overline{S} = \overline{MD_2 \cdot 1 \cdot 65S}$ + 2 Na	4.78	?	NaA 0.25 , NaD<10^{-3}	S<0.01 in benzene	66W
	D^- = Dinonyl naphthalene sulphonate					
Ligand: I^- , A^-						
Li^+	$M^+ + A^- + 2H_2O + 3\overline{S} = \overline{MA \cdot 2H_2O.3S}$	-0.66	?	MA 0.1 - sat.	S 0.5 in benzene	67R
	$M^+ + A^- + 2H_2O = \overline{MA \cdot 2H_2O}$	-1.55	?			
Na^+	$M^+ + A^- + 2.7\ H_2O + 2\ \overline{S} = \overline{MA \cdot 2.7\ H_2O.2S}$	-1.96	?			
	$M^+ + A^- + 2.7\ H_2O = \overline{MA \cdot 2.7\ H_2O}$	-2.57	?			
K^+	$M^+ + A^- + 5\ H_2O = \overline{MA \cdot 5\ H_2O}$	-2.80	?			

References

58B Blake, C. A., Baes, C. F., Jr., Brown, K. B., Coleman, C. F. and White, S. C., Proc. Conf. Peaceful Uses Atom. En., 28, 289 (1958)

60B Boyd, G. E. and Larson, Q. V., J. Phys. Chem., 64, 988 (1960)

60Z Zingaro, R. A. and White, J. C., J. Inorg. Nucl. Chem., 12, 315 (1960)

61M Martin, B., Ockenden, D. W. and Foreman, J. K., J. Inorg. Nucl. Chem., 21, 96 (1961)

62B Baker, H. I., and Baes, C. F., Jr., J. Inorg. Nucl. Chem., 24, 1277 (1962)

62I Ishimori, T., Rhee, C. T. and Fujino, Nippon Genshiryoku Gakkaishi, 4, 837 (1962)

63I Ihle, H., Michael, H. and Murrenhoff, A., J. Inorg. Nucl. Chem., 25, 734 (1963); German Report JUL-74-PC (1962)

63S Soman, Y. D., Univ. Microfilms (Ann Arbor, Michigan, USA), order No. 63-6650 (1963)

63Z Zangen, M., J. Inorg. Nucl. Chem., 25, 581 (1963)

64A Alian, A., Proc. Conf. Peaceful Uses Atom. En. 12, 256 (1964)

64D Duyckaerts, D. and Goffart, J., EURATOM Report EUR 437 (1964)

64H Heyn, A. H. A. and Soman, Y. D., J. Inorg. Nucl. Chem., 26, 287 (1964)

64M Mrochek, J. E. and Banks, C. V., J. Inorg. Nucl. Chem., 25, 955 (1963); USAEC Report IS-827 (1964)

64P Parker, J. R. and Banks, C. V., USAEC Report IS-942 (1964)

64R Rozen, A. M. and Nikolotova, Z. I., Zh. Neorg. Khim., 9, 1725 (1964)

64V Varga, L. P. and Hume, D. N., Inorg. Chem., 3, 77 (1964)

65A Alian, A. and Moustapha, Z., Indian J. Chem., 3, 390 (1965)

65C Conocchioli, T. S., Tocher, M. I. and Diamond, R. M., J. Phys. Chem., 69, 1106 (1965)

65Ca Caligara, F., Dreze, Ph., Duyckaerts, G., Fuger, J., Goffart, J. and Müller, W., EURATOM Report EUR 2297 (1965)

65P Parker, J. R. and Banks, C. V., J. Inorg. Nucl. Chem., 27, 631 (1965)

65R Rozen, A. M., Nikolotova, Z. I., Petrov, K. A., Skotnikov, A. S. and Teterin, E. G., Radiokhimiya, 7, 517 (1965)

65V Varga, L. P., Wakley, W. D., Nicolson, L. S., Madden, M. L. and Patterson, J., Anal. Chem., 37, 1003 (1965)

66L Liem, D. H. and Dyrssen, D., Acta Chem. Scand., 20, 272 (1966)

66W Wang, S. M. and Li, N. C., J. Inorg. Nucl. Chem., 28, 1091 (1966)

67G Goffart, J. and Duyckaerts, G., Anal. Chim. Acta, 38, 529 (1967)

67L Liem, D. H., in "Solvent Extraction Chemistry", Eds. Dyrssen, D., Liljenzin, J. O., Rydberg, J., North-Holland, Amsterdam, p. 264 (1967)

67R Rozen, A. M. and Mikhailichenko, A. I., Zhur. Neorg. Khim., 12, 743 (1967)

68K Krasovec, F. and Klofutar, C., J. Inorg. Nucl. Chem., 30, 1640 (1968)

68L Liem, D. H., Acta Chem. Scand., 22, 753 (1968)

METAL ION, M	REACTION	LOG K	TEMP.	CONDITIONS, AQ.	CONDITIONS, ORG.	REF.
Extractant: Triphenylphosphine oxide (TPh'PO)						
Reactions with acidic organophosphorus compounds						
	$HS' + \overline{S} = \overline{HS'.S}$	3.72	?	$HClO_4$ 0.1	S<0.1 in CCl_4	67K
	HS' = di(p-chlorophenyl)phosphoric acid					
Reactions with diluents						
	$\overline{ROH} + \overline{S} = \overline{ROHS}$	3.02	20	——	ROH<0.006, S in CCl_4	60A
	ROH = phenol	2.56	50			
	ROH = methanol	1.38	20	——	ROH<0.006, S<0.06 in CCl_4	61G
		1.02	50			
	ROH = α naphthol	3.06	20	——	ROH<0.006, S<0.06 in CCl_4	
		2.58	50			
	ROH = pentachlorophenol	2.83	20	——	ROH<0.006, S<0.06 in CCl_4	
		2.43	50			
Ligand: TcO_4^- , MO_4^-						
H^+	$MO_4^- + H^+ + \overline{S} = \overline{S.HMO_4}$	1.20	?	HCl 1.0	S<0.1 in dichloromethane	67K

METAL ION, M	REACTION	LOG K	TEMP.	CONDITIONS, AQ.	CONDITIONS, ORG.	REF.
Ligand: ReO_4^- , MO_4^-						
H^+	$MO_4^- + H^+ + \overline{S} = \overline{S.HMO_4}$	0.72	?	HCl 1.0	S<0.1 in dichloromethane	67K
Ligand: $SO_4^{=}$, $A^{=}$						
H^+	$H^+ + S = SH^+$	-2.70	?	H_2A , S 0.1	——	67K
Ligand: Cl^- , A^-						
UO_2^{++}	$M^{++} + 2A^- + \overline{S} = \overline{S.MA_2}$	2.54	?	HA 1.0	S<0.1 in dichloromethane	67K

References

60A Aksnes, G. and Gramstad, T., Acta Chem. Scand., 14, 1485 (1960)

61G Gramstad, T., Acta Chem. Scand. 15, 1337 (1961)

67K Krasovec, F. and Klofutar, C., in "Solvent Extraction Chemistry", Eds. Dyrssen, D., Liljenzin, J. O., Rydberg, J., North-Holland, Amsterdam, p. 509 (1967)

METAL ION, M	REACTION	LOG K	TEMP.	CONDITIONS, AQ.	CONDITIONS, ORG.	REF.
Extractant: Methylene bis(dihexyl phosphine oxide), B(DHx'PO)M'						
Ligand: NO_3^- , A^-						
H^+	$H^+ + A^- + \overline{S} = \overline{HA.S}$	0.51	25	HA 0.3	S<0.05 in 1.2 dichlorobenzene	64M
		1.18	25	HA 1	S<0.05 in 1.2 dichlorobenzene	
		0.99	25	HA 0.01 , NaA 2	S<0.1 in benzene	67G
	$H^+ + A^- + \overline{HA.S} = \overline{(2HA).S}$	-0.56	25			
Th^{+4}	$M^{+4} + 4A^- + \overline{S} = \overline{MA_4S}$	2.69	25	MA_4 0.1 , HA 3.10^{-3}	S<8.10^{-3} in 1.2 dichloro-benzene	64P,65P
	$M^{+4} + 4A^- + 2\overline{S} = \overline{MA_4.2S}$	9.73	25	HA 0.1	S<0.015 in 1.2 dichloro-benzene	
UO_2^{++}	$M^{+2} + 2A^- + 2\overline{S} = \overline{MA_2.2S}$	8.4	25	HA 0.013	S<0.025 in 1.2 dichloro-benzene	64M
		6.0	25	HA 1	S<0.1 in 1.2 dichlorobenzene	
		3.3	25	MA 0.1	S 0.01 in 1.2 dichlorobenzene	
NpO_2^{++}		3.9	25	HA 1.0	S<0.1 in 1.2 dichlorobenzene	67A
Ligand: ClO_4^- , A^-						
Ca^{+2}		5.4-6.6	?	NaA<3	S 0.1 in 1 2 dichlorobenzene	63W
NpO_2^{++}		3.3-3.9	25	HA 1.35	S<0.1 in 1.2 dichlorobenzene	67A
PuO_2^{++}		5.3	25			

References

63W Ferguson, J. W. and Banks, C. V., USAEC Report IS-777 (1963)

64M Mrochek, J. E. and Banks, C. V., USAEC Report IS-827 (1964); J. Inorg. Nucl. Chem., 25, 955 (1963)

64P Parker, J. R., and Banks, C. V., J. Inorg. Nucl. Chem., 25, 955 (1963); and USAEC Report IS-827 (1964)

65P Parker, J. R., and Banks, C. V., J. Inorg. Nucl. Chem., 27, 631 (1965)

67A Aziz, A., Univ. of Utah, Ph. D. Thesis, 1967, Univ. Microfilms

67G Goffart, J. and Duyckaerts, G., Anal. Chim. Acta, 39, 57 (1967)